AF787250

Numerical Modeling and Computer Simulation

Edited by Dragan M. Cvetković and Gunvant A. Birajdar

Published in London, United Kingdom

IntechOpen

Supporting open minds since 2005

Numerical Modeling and Computer Simulation
http://dx.doi.org/10.5772/intechopen.77641
Edited by Dragan M. Cvetković and Gunvant A. Birajdar

Contributors
Ghazala Ashraf, Khalid Saifullah Syed, Muhammad Ishaq, Stephen Kirkup, Javad Yazdani, Babacar Leye, Léon Matar Tine, Mamadou Sy, Marwa Ben Said-Romdhane, Sondes Skander-Mustapha, Ilhem Slama-Belkhodja, Akhilanand Chaurasia, Saman Ishrat, M. Husain Khan, Bakrim Fadwa, El Maroufy Hamid, Ait Mousse Hassan, Goharik Petrosyan

© The Editor(s) and the Author(s) 2020
The rights of the editor(s) and the author(s) have been asserted in accordance with the Copyright, Designs and Patents Act 1988. All rights to the book as a whole are reserved by INTECHOPEN LIMITED. The book as a whole (compilation) cannot be reproduced, distributed or used for commercial or non-commercial purposes without INTECHOPEN LIMITED's written permission. Enquiries concerning the use of the book should be directed to INTECHOPEN LIMITED rights and permissions department (permissions@intechopen.com).
Violations are liable to prosecution under the governing Copyright Law.

CC BY

Individual chapters of this publication are distributed under the terms of the Creative Commons Attribution 3.0 Unported License which permits commercial use, distribution and reproduction of the individual chapters, provided the original author(s) and source publication are appropriately acknowledged. If so indicated, certain images may not be included under the Creative Commons license. In such cases users will need to obtain permission from the license holder to reproduce the material. More details and guidelines concerning content reuse and adaptation can be found at http://www.intechopen.com/copyright-policy.html.

Notice
Statements and opinions expressed in the chapters are these of the individual contributors and not necessarily those of the editors or publisher. No responsibility is accepted for the accuracy of information contained in the published chapters. The publisher assumes no responsibility for any damage or injury to persons or property arising out of the use of any materials, instructions, methods or ideas contained in the book.

First published in London, United Kingdom, 2020 by IntechOpen
IntechOpen is the global imprint of INTECHOPEN LIMITED, registered in England and Wales, registration number: 11086078, 7th floor, 10 Lower Thames Street, London, EC3R 6AF, United Kingdom
Printed in Croatia

British Library Cataloguing-in-Publication Data
A catalogue record for this book is available from the British Library

Additional hard and PDF copies can be obtained from orders@intechopen.com

Numerical Modeling and Computer Simulation
Edited by Dragan M. Cvetković and Gunvant A. Birajdar
p. cm.
Print ISBN 978-1-83881-196-9
Online ISBN 978-1-83881-197-6
eBook (PDF) ISBN 978-1-83881-198-3

For EU product safety concerns:
IN TECH d.o.o., Prolaz Marije Krucifikse Kozulić 3, 51000 Rijeka, Croatia, info@intechopen.com or visit our website at intechopen.com.

We are IntechOpen, the world's leading publisher of Open Access books

Built by scientists, for scientists

4,800+
Open access books available

122,000+
International authors and editors

135M+
Downloads

151
Countries delivered to

Our authors are among the
Top 1%
most cited scientists

12.2%
Contributors from top 500 universities

CLARIVATE ANALYTICS
BOOK CITATION INDEX
INDEXED

WEB OF SCIENCE™

Selection of our books indexed in the Book Citation Index
in Web of Science™ Core Collection (BKCI)

Interested in publishing with us?
Contact book.department@intechopen.com

Numbers displayed above are based on latest data collected.
For more information visit www.intechopen.com

Meet the editors

Dragan Cvetković graduated in Aeronautics from the Faculty of Mechanical Engineering, University of Belgrade, in 1988. In the Aeronautical Department, he defended his doctoral dissertation in 1997. So far he has published 64 books, scripts and practicums about computers and computer programs, the aviation weapons and flight mechanics. He has published a large number of scientific papers in his country as well as abroad. Since 2007, he has been working at the SINGIDUNUM UNIVERSITY in Belgrade as an assistant professor, and since 2019 as the Vice Rector for Teaching. He became a Full Professor in the field of Informatics and Computing in 2014.

Gunvant Birajdar is a post-graduate in Mathematics from the Department of Mathematics, Dr. Babasaheb Ambedkar Marathwada University Aurangabad Faculty of Science (2008). He completed a thematic semester on "Numerical Computing for Algebraic Geometry" at Max-Planck Institute for Mathematical Sciences, Germany in 2018. He also completed a thematic semester on "Fundamental Practice of Finite Element Method"at University of Rennes 1, France in 2018. He was a Doctoral fellow under Research Fellowship in Sciences for Meritorious Students (RFSMS) UGC, New Delhi from 2009 to 2013. He published more than 15 research articles in internationally reputed journals. He is an editorial member of several international journals as well as reviewer for many journals. He received different awards for his contribution in the research. Currently he is working at School of Rural Development, Tata Institute of Social Sciences Tuljapur, Dist. Osmanabad, Maharashtra, India.

Contents

Preface

Information technologies have changed people's lives to a great extent, and now it is almost impossible to imagine any activity that does not depend on computers in some way. Since the invention of first computer systems, people have been trying to avail computers in order to solve complex problems in various areas. The need for computer systems in the calculation of different building and automotive constructions has appeared with the development of industry. Traditional methods of calculation have been replaced by computer programs that have the ability to predict the behavior of structures under different loading conditions. Hence, expensive experiments, tests and examinations had been replaced by cheaper and more powerful computational methods that do not require the destruction of structure in order to determine its capacity.

Computer simulation or a computer model has the task to simulate the behavior of an abstract model of a particular system. Computer simulations have become a useful part of mathematical modeling of many natural systems in physics, quantum mechanics, chemistry and biology, then in economic systems, psychology and social sciences, as well as in the engineering process of new technologies, in order to gain a better insight into their way of working and behaving.

In comparison to classical modeling (model making, prototyping, etc.), mathematical and numerical modeling is more cost-effective and model itself can be checked and examined much faster and in a more efficient way, therefore one does not need to use considerable material resources - money, raw materials for modeling, energy, etc. Mathematical modeling provides more freedom for engineers to utilize different variations of components, elements and parameters of conceived technical solutions. The advantage of this type of modeling is the fact that only important and poorly tested elements of the system can be submitted to modeling.

Any kind of previously mentioned process of modeling requires the engineers' excellent knowledge in the field one is considering, as well as the methods used for solving the required task. The attained knowledge through permanent education has the significant influence on the quality of designed model.

Computer simulations are different from computer programs. Contrary to computer programs that run for a few minutes, simulations can be run on the local network and can last for hours or even for days. It can be said that the simulations exceeded many methods that use pencil and paper for solving any kind of mathematical problems.

The first chapter of this book illustrates the usage of pilot Fortran software library for the solution of Laplace's Equation by the boundary element method (BEM). The second chapter describes the optimal control of thermal pollution emitted by power plants. In the third chapter, the authors present the finite difference solution of conjugate heat transfer in double pipe with trapezoidal fins. The fourth chapter deals with a photovoltaic system integrated into the buildings. The fifth chapter presents the way in which computer simulation affect the practice of diagnosis in the field of oral medicine and radiology. In the sixth chapter, the authors write about the simulation and parametric inference of a mixed effects model with stochastic differential equations using the Fokker – Planck equation solution. The seventh chapter illustrates the possibilities of modeling Petri nets and their extensions. Finally, the eighth chapter presents modeling, control design, simulation and experimental validation of the line impedance emulator.

We would like to express our sincere gratitude to all the authors and coauthors for their contribution. The successful completion of the book *Numerical Modeling and Computer Simulation* has been the result of the cooperation between many people. We would especially like to thank the Publishing Process Manager Ms. Dajana Pemac for her support during the publishing process.

Dragan M. Cvetković
Singidunum University,
Faculty of Informatics and Computing,
Belgrade, Republic of Serbia

Gunvant A. Birajdar
Tata Institute of Social Sciences,
School of Rural Development,
Tuljapur, District - Osmanabad, Maharashtra

Section 1

Numerical Modeling

Chapter 1

A Pilot Fortran Software Library for the Solution of Laplace's Equation by the Boundary Element Method

Stephen Kirkup and Javad Yazdani

Abstract

The boundary element method (BEM) is developed from the standpoint of software design. The Fortran language is used to produce a structured library for solving Laplace's equation in various domain topologies and dimensions with generalised boundary conditions. Subroutines that compute the discrete Laplace operators, which are the core components for populating the matrices in the BEM, are developed. The main subroutines for solving Laplace's equation in 2D, 3D and axisymmetric cases for open and closed boundaries are introduced. The methods are demonstrated on test problems.

Keywords: boundary element method, Laplace's equation, Fortran

1. Introduction

The boundary element method (BEM) has established itself as an important numerical technique for solving partial differential equations (PDEs) over the last half century [1, 2]. It distinguishes itself from competing methods, such as the finite element method (FEM) [3] in that the latter method requires a mesh of the domain, whereas the BEM only requires a mesh of the boundary (of the domain). The BEM is not as widely applicable as the FEM, particularly in that it is much more of a struggle to apply the BEM to non-linear problems. However, for problems to which the boundary element method is viable, the advantage of only requiring a boundary mesh is a significant one; the BEM is likely to be more efficient but also the relative simplicity of meshing, and the method is easier to use and is more accessible. This advantage is more notable for exterior problems; the domain is infinite, and 'domain methods' such as the FEM require special treatment, but for the BEM, only a (finite) boundary mesh is required. Computational methods may be combined or coupled [2].

The boundary element method is derived through the discretisation of an integral equation that is mathematically equivalent to the original partial differential equation. The essential reformulation of the PDE that underlies the BEM consists of an integral equation that is defined on the boundary of the domain and an integral that relates the boundary solution to the solution at points in the domain. The former is termed a boundary integral equation (BIE), and the BEM is often referred

IntechOpen

to as the boundary integral equation method or boundary integral method. There are two classes of boundary element method, termed the direct and indirect method. The direct method is based on Green's second theorem, whereas the indirect method is based on describing the solution in terms of layer potentials. In this work the direct boundary element method is developed.

The simplest partial differential equation that is amenable to the BEM is Laplace's equation:

$$\sum_{i=1}^{N} \frac{\partial^2 \varphi(\boldsymbol{p})}{\partial x_i^2} = 0 \tag{1}$$

where N is the dimension of the space or, more concisely,

$$\nabla^2 \varphi = 0. \tag{2}$$

Laplace's equation therefore acts as a model problem for developing the BEM. Laplace's equation also has a number of applications; steady-state heat conduction, steady-state electric potential, gravitation and groundwater flow [4–13].

Initially, in this paper, the derivation of the direct boundary element method is introduced for the interior two-dimensional Laplace problem. The boundary element method is developed in Fortran for the 2D Laplace problem; then this is extended to axisymmetric three-dimensional problems and to both interior and exterior problems. The boundary element method can be extended to problems where the body being modelled is 'thin', like a screen or discontinuity, and these are also included. Test problems are applied to the codes, and the results are given for all problem classes. There are a number of studies on numerical error in the boundary element method [14–16].

There have been a number of works on coding the boundary element method [17–19]. The focus of this work is the algorithms and the software for solving Laplace problems by the BEM. As with the earlier works by the first author on Laplace and Helmholtz (acoustic) problems [20–24], this is about continuing with the development of a base library of methods and corresponding software. The codes and guides can be found on the first author's website [25].

The codes have been developed in Fortran 77, but the language is just used to provide a simple template for exploring the methods and the organisation of coding. The algorithms and coding for Laplace's equation considered in this work also provide a useful basis for the development of the BEM for other problems and add to the library of numerical software [26].

2. The BEM and the 2D interior Laplace problem

The Laplace equation provides a useful model problem for the boundary element method. The two-dimensional case is the simplest of these and is the best place to start to learn about the method. In this section the solution of Laplace's equation in an interior domain by the direct BEM is outlined, and this also provides the foundation for the 3D BEM development in later sections.

2.1 Boundary integral equation formulation of the interior Laplace problem

Laplace's equation (2) governs the interior domain D enclosed by a boundary S. The solution must also satisfy a boundary condition, and it is important in terms of maintaining the generality of the method that this is in a general (Robin) form:

$$\alpha(\boldsymbol{p})\boldsymbol{\varphi}(\boldsymbol{p}) + \beta(\boldsymbol{p})\frac{\partial \varphi}{\partial n_p}(\boldsymbol{p}) = f(\boldsymbol{p})(\boldsymbol{p} \in S). \tag{3}$$

In the direct BEM, Laplace's equation is replaced by an equivalent integral equation of the form:

$$\int_S \frac{\partial G(\boldsymbol{p},\boldsymbol{q})}{\partial n_q}\varphi(\boldsymbol{q})dS_q + \frac{1}{2}\varphi(\boldsymbol{q}) = \int_S G(\boldsymbol{p},\boldsymbol{q})\frac{\partial \varphi(\boldsymbol{q})}{\partial n_q}dS_q(\boldsymbol{p} \in S), \tag{4}$$

$$\int_S \frac{\partial G(\boldsymbol{p},\boldsymbol{q})}{\partial n_q}\varphi(\boldsymbol{q})dS_q + \varphi(\boldsymbol{q}) = \int_S G(\boldsymbol{p},\boldsymbol{q})\frac{\partial \varphi(\boldsymbol{q})}{\partial n_q}dS_q(\boldsymbol{p} \in D). \tag{5}$$

The terminology $\frac{\partial *}{\partial n_q}$ represents the partial derivative of the function* with respect to the unit outward normal at point $\boldsymbol{q}$ on the boundary. The function G is known as Green's function. Physically, $G(\boldsymbol{p}, \boldsymbol{q})$ represents the effect observed at point $\boldsymbol{p}$ of a unit source at point $\boldsymbol{q}$. For the Laplace equation, Green's function is denoted by G and is defined as

$$G(\mathbf{p},\mathbf{q}) = -\frac{1}{2\pi}\ln r \tag{6}$$

for two-dimensional Laplace problems, where $r = |\boldsymbol{q} - \boldsymbol{p}|$.

Integral Eqs. (4) and (5) can be derived from the Laplace equation by applying Green's second theorem. The power of the formulation lies in the fact that Eq. (4) relates the potential φ and its derivative on the boundary alone; no reference is made to φ at points in the domain in this particular boundary integral equation. In a typical boundary value problem, we may be given $\varphi(\boldsymbol{q})$, $\frac{\partial \varphi(\boldsymbol{q})}{\partial n_q}$ or a combination of such data on S. The boundary integral equation is a means of determining the unknown boundary function(s), followed by the domain solution from the given boundary data.

2.2 Operator notation

Operator notation is a useful shorthand in writing integral equations. Moreover, it will be shown that it is a very powerful notation in that it clearly demonstrates the connection between the integral equation and the linear system of equations that results from its discretisation.

Integral equations can always be written in terms of integral operators. For example, if ζ is a function defined on a (closed or open) boundary Γ, then applying the following operation to ζ for all points $\boldsymbol{p}$ on Γ

$$\int_\Gamma G(\boldsymbol{p},\boldsymbol{q})\zeta(\boldsymbol{q})dS_q = \mu(\boldsymbol{p})(\boldsymbol{p} \in S) \tag{7}$$

gives a function μ. This may be viewed as the application of an operator to the function ζ to return the function μ. More simply we may write

$$\{L\zeta\}_\Gamma(\mathbf{p}) = \mu(\mathbf{p}). \tag{8}$$

In Eq. (8) L represents the integral operator, and the subscript (Γ) refers to the domain of integration. Γ is used as a variable, representing either a whole boundary or a part of the boundary. The other three important Laplace integral operators are defined as follows:

$$\{M\zeta\}_\Gamma(\boldsymbol{p}) = \int_\Gamma \frac{\partial G(\boldsymbol{p},\boldsymbol{q})}{\partial n_q}\zeta(\boldsymbol{q})dS_q, \tag{9}$$

$$\{M^t\zeta\}_\Gamma(\boldsymbol{p};\boldsymbol{v}_p) = \frac{\partial}{\partial v_p}\int_\Gamma G(\boldsymbol{p},\boldsymbol{q})\zeta(\boldsymbol{q})dS_q, \tag{10}$$

$$\{N\zeta\}_\Gamma(\boldsymbol{p};\boldsymbol{v}_p) = \frac{\partial}{\partial v_p}\int_\Gamma \frac{\partial G(\boldsymbol{p},\boldsymbol{q})}{\partial n_q}\zeta(\boldsymbol{q})dS_q, \tag{11}$$

where $\boldsymbol{v}_p$ is any unit vector. In operator notation of the previous subsection, the integral equation formulation (3) can be written in the following form:

$$\left\{\left(M+\frac{1}{2}I\right)\varphi\right\}_S(\boldsymbol{p}) = \{Lv\}_S(\boldsymbol{p})(\boldsymbol{p}\in S), \tag{12}$$

$$\varphi(\boldsymbol{p}) = \{Lv\}_S - \{M\varphi\}_S(\boldsymbol{p}\in D), \tag{13}$$

where $v(\boldsymbol{q}) = \frac{\partial\varphi(\boldsymbol{q})}{\partial n_q}$.

2.3 Direct boundary element method

For the direct boundary element method solution of the interior Laplace problem, that is, developed in this section, the initial stage involves solving boundary integral Eq. (4), returning (approximations to) both φ and ∂φ/∂n on S. The second stage of the BEM involved finding the solution at any chosen points in the domain D. The most straightforward method for solving integral equations like Eq. (4) is that of collocation. Collocation may be applied in a remarkably elementary form, which is termed C^{-1} collocation in this text since it is derived by approximating the boundary functions by a constant on each panel. In this subsection the C^{-1} collocation method is briefly outlined.

To begin with, the boundary S is assumed to be expressed as a set of panels:

$$S \approx \tilde{S} = \sum_{j=1}^{n} \Delta\tilde{S}_j. \tag{14}$$

Usually the panels have a characteristic form and cannot represent a given boundary exactly. For example, a two-dimensional boundary can be approximated by a set of straight lines. In order to complete the discretisation of the integral equations, the boundary functions also need to be approximated on each panel. In this work, it is the characteristics of the panel and the representation of the boundary function on the panel that together define the element in the boundary element method. By representing the boundary functions by a characteristic form on each panel, the boundary integral equations can be written as a linear system of equations of the form introduced earlier.

The term element refers not only to the form of ΔS_j but also to the method of representing the boundary functions on ΔS_j. The C^{-1} collocation method involves representing the boundary function by a constant on each panel:

$$\varphi(\boldsymbol{p}) \approx \varphi_j, v(\boldsymbol{p}) \approx v_j\left(\boldsymbol{p}\in\Delta\tilde{S}_j\right). \tag{15}$$

The substitution of representations of this form for the boundary functions in the integral equation reduces it to discrete form. The simplifications allow us to rewrite Eq. (11) as the approximation:

$$\sum_{j=1}^{n}\left\{\left(M+\frac{1}{2}I\right)e\right\}_{\Delta\tilde{S}_j}(\boldsymbol{p})\varphi_j \approx \sum_{j=1}^{n}\{\mathrm{Le}\}_{\Delta\tilde{S}_j}(\boldsymbol{p})v_j\ \left(\boldsymbol{p}\in\tilde{S}\right), \tag{16}$$

where e is the unit function ($e \equiv 1$). $\{Le\}_{\Delta\tilde{S}_j}(\boldsymbol{p})$, for example, for a specific point $\boldsymbol{p}$, are the numerical values of definite integrals that together can be interpreted as the discrete form of the L integral operator.

The constant approximation is taken to be the value of the boundary functions at the representative central point (the collocation point) on each panel. By finding the discrete forms of the relevant integral operators for all the collocation points, a system of the form

$$\sum_{j=1}^{n}\left\{\left(M+\frac{1}{2}I\right)e\right\}_{\Delta\tilde{S}_j}(\boldsymbol{p}_{Si})\varphi_j \approx \sum_{j=1}^{n}\{Le\}_{\Delta\tilde{S}_j}(\boldsymbol{p}_{Si})v_j\ \left(\boldsymbol{p}_{Si}\in\tilde{S}\right), \tag{17}$$

for i = 1, 2 and n is obtained by putting $\boldsymbol{p} = \boldsymbol{p}_{Si}$ in the previous approximation. Note that because of the approximation of the boundary functions (and also the boundary approximation, if applicable), the discrete equivalent of Eq. (12) is an approximation relating the exact values of the boundary functions at the collocation points.

This system of approximations can now be written in the matrix-vector form:

$$\left(M_{SS}+\frac{1}{2}\mathrm{I}\right)\underline{\hat{\varphi}}_S = L_{SS}\underline{\hat{v}}_S, \tag{18}$$

with the matrix components defined by $[L_{SS}]_{ij} = \{Le\}_{\Delta\tilde{S}_j}(\boldsymbol{p}_{Si})$, $[M_{SS}]_{ij} = \{Me\}_{\Delta\tilde{S}_j}(\boldsymbol{p}_{Si})$. The vectors $\underline{\hat{\varphi}}_S$ and $\underline{\hat{v}}_S$ are representative or approximate values of φ and v at the collocation points. In the first stage of the boundary element, the system (18) is solved alongside the discrete form of the boundary condition (3):

$$\alpha_i\varphi_i + \beta_i v_i = f_i \text{ for } i = 1, 2, \ldots n. \tag{19}$$

The discrete forms are definite integrals that need to be computed usually by numerical integration. For the solution of Eqs. (18) and (19), the (approximation to) boundary data is known at the collocation points.

Once the (approximations to) functions on the boundary are known, after completing the initial stage of the boundary element method, the domain solution can be found. In the case of the interior Laplace problem, Eq. (13) will yield the domain solution. Similarly, the discrete equivalent of Eq. (11) may be derived:

$$\varphi(\boldsymbol{p}_{Di}) = \sum_{j=1}^{n}\{Le\}_{\Delta\tilde{S}_j}(\boldsymbol{p}_{Di})\hat{v}_j - \sum_{j=1}^{n}\{Me\}_{\Delta\tilde{S}_j}(\boldsymbol{p}_{Di})\hat{\varphi}_j\left(\boldsymbol{p}_{Di}\in\tilde{D}\right), \tag{20}$$

for each point $\boldsymbol{p}_{Di}$ in the domain $\tilde{D}$. Let the solution be sought at m domain points $\boldsymbol{p}_{Di}$ for $i = 1, 2, \ldots m$, and then the equation above, for all the domain points, is written as

$$\underline{\hat{\varphi}}_D = L_{DS}\underline{\hat{v}}_S - M_{DS}\underline{\hat{\varphi}}_S, \tag{21}$$

where $[L_{DS}]_{ij} = \{Le\}_{\Delta\tilde{S}_j}(\boldsymbol{p}_{Di})$, $[M_{DS}]_{ij} = \{Me\}_{\Delta\tilde{S}_j}(\boldsymbol{p}_{Di})$.

2.4 LIBEM2 and L2LC modules, test problem and results

This section includes an outline of the Laplace interior BEM 2D (LIBEM2) and Laplace 2D linear boundary approximation constant element (L2LC) modules, and they are demonstrated by means of a test problem. The LIBEM2 module solves Laplace's equation in an interior two-dimensional domain. L2LC is the most important component module.

L2LC. The L2LC module computes the discrete Laplace operators for two-dimensional problems. In the notation of this article, the routine computes $\{Le\}_{\Delta\tilde{S}_j}(\boldsymbol{p})$, $\{Me\}_{\Delta\tilde{S}_j}(\boldsymbol{p})$, $\{M^te\}_{\Delta\tilde{S}_j}(\boldsymbol{p})$ and $\{Ne\}_{\Delta\tilde{S}_j}(\boldsymbol{p})$, where $\Delta\widetilde{S}_j$ is the panel that is the domain of integration and $\boldsymbol{p}$ is any point. The call to the subroutine has the following form:

```
SUBROUTINE L2LC(P,VECP,QA,QB,LPONEL, LVALID,EGEOM,LFAIL,
     * NEEDL,NEEDM,NEEDMT,NEEDN,L0,M0,M0T,N0),
```

where P is point $\boldsymbol{p}$; VECP is a unit directional vector that passes through $\boldsymbol{p}$; QA and QB are the points, either side of the panel and hence defining the panel; LPONEL is a logical switch that declares whether $\boldsymbol{p}$ is on the panel; and NEEDL is a logical switch that states whether the discrete L operator is required and similar to the other operators. The computed values for the integrals are output in L0, M0, M0T and N0. For the straightforward direct BEM, developed in the previous section, only L and M operators are required.

In general L2LC simply implements a Gaussian quadrature rule in order to determine the integral, using a higher-order rule when point $\boldsymbol{p}$ is close to the panel $\Delta\widetilde{S}_j$. However, when point $\boldsymbol{p}$ is on the panel, then an exact integration is used [21, 22].

LIBEM2. The LIBEM2 module solves the interior Laplace problem and has the following form:

```
LIBEM2(MAXNODES,NNODE,NODES,MAXPANELS,NPANEL,PANELS,
     * MAXPOINTS,NPOINT,POINTS,
     * SALPHA,SBETA,SF,SINPHI,PINPHI,
     * LSOL,LVALID,TOLGEOM,
     * SPHI,SVEL,PPHI,
     * L_SS,M_SSPMHALFI,L_PS,M_PS,
     * PERM,XORY,C,workspace)
```

The boundary is set up through listing a set of nodal coordinates, and each panel is determined through the two nodal indices for the endpoints of the panel. The nodal coordinates are input through NODES and the panel information through PANELS. The nodes are oriented clockwise on each panel for an outer boundary and anticlockwise for any inner boundary. Usually, a solution in the domain is sought, and for this a set of (interior) domain points are set in POINTS. The boundary condition is set with the parameters SALPHA, SBETA and SF, setting α_i, β_i and f_i values in Eq. (19) for each panel.

Test problem. The test problem is that of solving Laplace's equation on a unit square with the boundary conditions defined as shown in **Figure 1**. The solution is sought at the five interior points (0.25, 0.25), (0.75, 0.25), (0.25, 0.75), (0.75, 0.75) and (0.5, 0.5), and these are also illustrated in the figure.

The test problem is set up in the file LIBEM2_T.FOR. The boundary is defined by 32 nodes and panels. The nodes are indexed, starting with 1: (0.0, 0.0), 2: (0.0, 0.125), 3: (0.0, 0.25) and continue clockwise around the boundary until the final node 32: (0.125, 0.0). The panels are similarly set up in the clockwise sense with panel 1:1–2 (panel 1 links node 1 with node 2) and 2:2–3 until the final panel 32:32–1, linking the

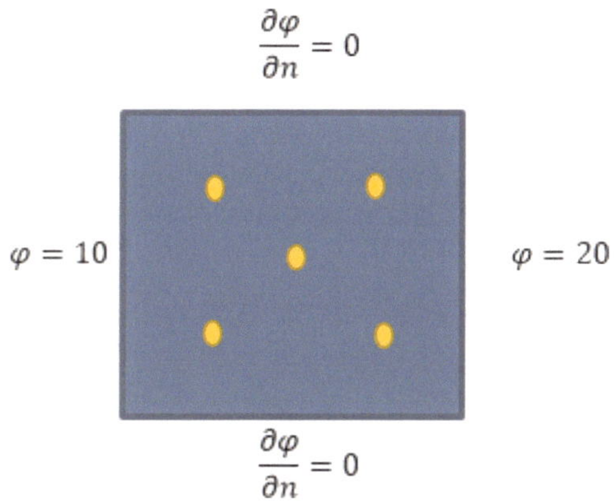

Figure 1.
The test problem of the unit square domain with boundary conditions.

Index	Point	Exact	Computed (5 d.p.)
1	(0.25, 0.25)	12.5	12.49568
2	(0.75, 0.25)	17.5	17.50432
3	(0.25, 0.75)	12.5	12.49568
4	(0.75, 0.75)	17.5	17.50432
5	(0.5, 0.5)	15.0	15.00000

Table 1.
The results from the two-dimensional interior problem.

final node with the first node to complete the boundary. The boundary conditions shown in **Figure 1** are then applied.

The exact solution is $\varphi(\boldsymbol{p}) = 10 + 10x$; this is clearly a solution of Laplace's equation and satisfies the boundary conditions. The exact solution at the interior points is therefore $\varphi = 12.5$ at the two points on the left, $\varphi = 17.5$ at the two points on the right and $\varphi = 15.0$ at the central point. The exact and computed results are shown in **Table 1**.

A set of nodal coordinates and each panel is determined through the two nodal indices for the endpoints of the panel.

The results from this test problem are also intuitively correct. With the left and right sides of the square at different potentials, it is common sense to expect the potential in the middle to be halfway between etc. The potentials can—most simply—be interpreted as temperatures in a steady-state heat conduction problem.

3. The BEM and 3D Laplace problems

In this section, the boundary element method—introduced for two-dimensional problems in the previous section—is extended to include three-dimensional problems in this section. In this section, the three-dimensional boundary may be general, but the special case of axisymmetric problems is also developed in the modules LBEM3 and LBEMA. The modules can solve interior and exterior Laplace problems.

For interior three-dimensional problems, the basic integral formulation is the same as for 2D problems (12) and (13), except that Green's function for three-dimensional Laplace problems is

$$G(\boldsymbol{p},\boldsymbol{q})=\frac{1}{4\pi r}, \tag{22}$$

where r is the distance between points $\boldsymbol{p}$ and $\boldsymbol{q}$ and the integrals are over surfaces rather than lines. The equations for the exterior problem are the same as for the interior problem, but for some changes of sign

$$\left\{\left(M-\frac{1}{2}I\right)\varphi\right\}_S(\boldsymbol{p})=\{Lv\}_S(\boldsymbol{p})(\boldsymbol{p}\in S), \tag{23}$$

$$\varphi(\boldsymbol{p})=\{M\varphi\}_S-\{Lv\}_S(\boldsymbol{p}\in E). \tag{24}$$

For general three-dimensional problems, the simplest elements are triangular panels, and for axisymmetric problems, they are lateral sections of a cone, with surface functions approximated by a constant on each panel.

3.1 LBEMA and L3ALC modules, test problems and results

Let us start on the introduction of three-dimensional problems with the axisymmetric codes. These codes are used in a very similar manner. As with the two-dimensional problem, the component module L3ALC computes the integrals over the panels and is called as follows:

```
SUBROUTINE L3ALC(P,VECP,QA,QB,LPONEL,LVALID,EGEOM,LFAIL,
     * NEEDL,NEEDM,NEEDMT,NEEDN,DISL,DISM,DISMT,DISN).
```

For axisymmetric problems, the surface is defined by conical panels, which are defined by piecewise straight lines along the generator. The parameters follow a similar pattern as L2LC, except the points and vectors are in cylindrical (r,z) coordinates. QA and QB are the two points either side of the panel on the generator.

The LBEMA subroutine computes the solution of the Laplace equation by the direct boundary element method and has the following form:

```
LBEMA(MAXNODES,NNODE,NODES,MAXPANELS,NPANEL,PANELS,
     * LINTERIOR,MAXPOINTS,NPOINT,POINTS,
     * SALPHA,SBETA,SF,SINPHI,PINPHI,
     * LSOL,LVALID,TOLGEOM,
     * SPHI,SVEL,PPHI,
     * L_SS,M_SSPMHALFI,L_PS,M_PS,
     * PERM,XORY,C,workspace)
```

In LBEMA, NODES lists the (r,z) coordinates of the nodes on the generator of the surface, and PANELS states the two nodes that together define each panel. LINTERIOR is a logical input, which is set to TRUE if an interior problem is to be solved and FALSE for an interior problem.

The interior test problem is in file LBEMA_IT. The test problem is the unit sphere with the exact solution:

$$\varphi=r^2-2z^2, \tag{25}$$

which is easily shown to be a solution of Laplace's equation by writing $r^2=x^2+y^2$. A Dirichlet boundary condition is applied; the solution is sought at four interior points, and the results for 18 elements are given in **Table 2**.

The exterior test problem is in file LBEMA_ET. The test problem is the unit sphere (approximated by 18 elements) with the exact solution:

Index	Point	Exact	Computed (4 d.p.)
1	(0.0, 0.0)	0.0	−0.0013
2	(0.0, 0.5)	−0.5	−0.4995
3	(0.0, −0.5)	−0.5	−0.4995
4	(0.5, 0.0)	0.25	0.2477

Table 2.
The results from the axisymmetric interior problem.

Index	Point	Exact (4 d.p.)	Computed (4 d.p.)
1	(0.0, 2.0)	0.5	0.4986
2	(1.0, 1.0)	0.7071	0.7051
3	(0.0, 100.0)	0.0100	0.0100

Table 3.
The results from the axisymmetric exterior problem.

$$\varphi = \frac{1}{r}, \tag{26}$$

where r is the distance from the origin. φ is a solution of Laplace's equation as it is a simple multiplication of Green's function (22). A Dirichlet boundary condition is applied to the upper hemisphere, and a Neumann boundary condition is applied on the lower hemisphere. The solution is sought at four interior points, and the results are given in **Table 3**.

3.2 LBEM3 and L3LC modules, test problems and results

The LBEM3 and L3LC subroutines implement the boundary element method for general three-dimensional problems. As with the two-dimensional and axisymmetric codes, the component module L3LC computes the integrals over the panels. The L3LC subroutine is called as follows:

```
SUBROUTINE L3LC(P,VECP,QA,QB,QC,LPONEL,LVALID,EGEOM,LFAIL,
   * NEEDL,NEEDM,NEEDMT,NEEDN,DISL,DISM,DISMT,DISN)
```

The parameters follow a similar purpose as did in the L2LC, except that the points and vectors have three values. QA, QB and QC are the coordinates of the vertices of the triangular panel.

The LBEM3 module solves Laplace's equation in a general interior or exterior three-dimensional domain and is called as follows:

```
LBEM3(MAXNODES,NNODE,NODES,MAXPANELS,NPANEL,PANELS,
   * LINTERIOR,MAXPOINTS,NPOINT,POINTS,
   * SALPHA,SBETA,SF,SINPHI,PINPHI,
   * LSOL,LVALID,TOLGEOM,
   * SPHI,SVEL,PPHI,
   * L_SS,M_SSPMHALFI,L_PS,M_PS,
   * PERM,XORY,C,WKSPC1,WKSPC2,WKSPC3)
```

As with LIBEM2 and LBEMA, NODES and PANELS define the boundary. However, in this case, NODES lists the three coordinates of each surface node, and PANELS lists the three nodal indices that make up each triangular panel.

Index	Point	Exact	Computed (4 d.p.)
1	(0.5, 0.0, 0.0)	0.5	0.4772
2	(0.0, 0.5, 0.0)	0.5	0.4836
3	(0.0, 0.0, 0.5)	0.5	0.4817
4	(0.1, 0.2, 0.3)	0.6	0.5802

Table 4.
The results from the three-dimensional interior problem.

Index	Point	Exact (4 d.p.)	Computed (4 d.p.)
1	(2.0, 0.0, 0.0)	0.4851	0.4969
2	(0.0, 4.0, 0.0)	0.2481	0.2536
3	(0.0, 0.0, 8.0)	0.1333	0.1360
4	(2.0, 2.0, 2.0)	0.3123	0.3189

Table 5.
The results from the three-dimensional exterior problem.

The interior test problem is that of a unit sphere approximated by 36 triangular panels. The exact solution that is applied as a Dirichlet boundary condition is

$$\varphi = x + y + z. \tag{27}$$

The results at four interior points are given in **Table 4**.

The exterior test problem is that of a unit sphere approximated by 36 triangular panels, as in the previous test. The exact solution that is applied as a Dirichlet boundary condition is

$$\varphi = \frac{1}{r}, \tag{28}$$

where r is the distance from $(0, 0, 0.5)$. The results at four exterior points are given in **Table 5**.

4. The solution of the 3D Laplace equation around a thin shell

Let us now consider the integral equation formulation for thin shells. An illustration of a typical problem of a hollow hemispherical cap is illustrated in **Figure 2**. In the traditional boundary element method, the boundaries are closed. This analysis and software design extends the boundary element method to open boundaries or discontinuities in the potential field.

In this section, the integral equations that are a reformulation of Laplace's equation surrounding a thin shell are stated. Fortran codes that implement the boundary element method for axisymmetric and general three-dimensional problems are outlined in this section and demonstrated on simple test problems, similar to the modelling of the steady-state electric field in a capacitor in Kirkup [9].

4.1 Integral equations and boundary element equations for thin shells

Following the work of Warham [27], the first step is to designate an 'upper' and 'lower' surface of a shell Γ and denote them by '+' and '−'. We then introduce the

DOI: http://dx.doi.org/10.5772/intechopen.86507

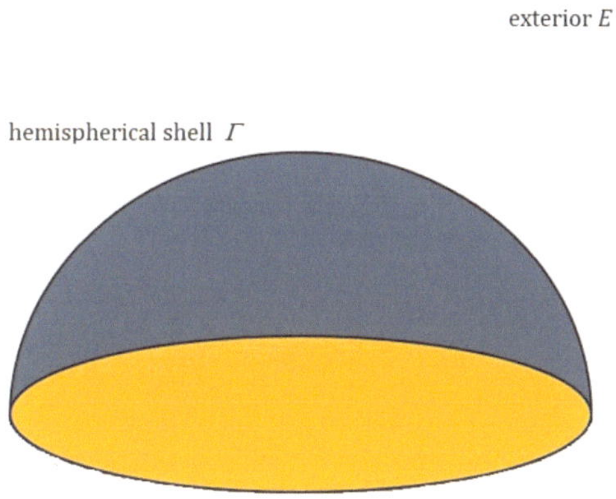

Figure 2.
A hemispherical shell.

quantities of difference and average of the potential and its normal derivative across the surface:

$$\delta(\boldsymbol{p}) = \varphi(\boldsymbol{p}_+) - \varphi(\boldsymbol{p}_+)(\boldsymbol{p} \in \Gamma), \tag{29}$$

$$\Phi(\boldsymbol{p}) = \frac{1}{2}(\varphi(\boldsymbol{p}_+) + \varphi(\boldsymbol{p}_+))(\boldsymbol{p} \in \Gamma), \tag{30}$$

$$\nu(\boldsymbol{p}) = v(\boldsymbol{p}_+) + v(\boldsymbol{p}_+)(\boldsymbol{p} \in \Gamma), \tag{31}$$

$$V(\boldsymbol{p}) = \frac{1}{2}(v(\boldsymbol{p}_+) - v(\boldsymbol{p}_+))(\boldsymbol{p} \in \Gamma). \tag{32}$$

The integral equation formulations for the Laplace equation in the exterior domain can now be written using the operator notation introduced earlier:

$$\varphi(\boldsymbol{p}) = \{M\delta\}_\Gamma(\boldsymbol{p}) - \{L\nu\}_\Gamma(\boldsymbol{p})(\boldsymbol{p} \in E), \tag{33}$$

$$\Phi(\boldsymbol{p}) = \{M\delta\}_\Gamma(\boldsymbol{p}) - \{L\nu\}_\Gamma(\boldsymbol{p})(\boldsymbol{p} \in \Gamma), \tag{34}$$

$$V(\boldsymbol{p}) = \{N\delta\}_\Gamma(\boldsymbol{p}) - \{M^t\nu\}_\Gamma(\boldsymbol{p})(\boldsymbol{p} \in \Gamma). \tag{35}$$

The boundary condition may be expressed in the following form:

$$\alpha(\boldsymbol{p})\boldsymbol{\delta}(\boldsymbol{p}) + \beta(\boldsymbol{p})\nu(\boldsymbol{p}) = f(\boldsymbol{p})(\boldsymbol{p} \in \Gamma), \tag{36}$$

$$A(\boldsymbol{p})\Phi(\boldsymbol{p}) + \beta(\boldsymbol{p})V(\boldsymbol{p}) = F(\boldsymbol{p})(\boldsymbol{p} \in \Gamma). \tag{37}$$

The discrete equivalents of Eq. (21) are as follows:

$$\underline{\hat{\varphi}}_E = M_{E\Gamma}\underline{\hat{\delta}}_\Gamma - L_{E\Gamma}\underline{\hat{\nu}}_\Gamma, \tag{38}$$

$$\underline{\hat{\Phi}}_\Gamma = M_{\Gamma\Gamma}\underline{\hat{\delta}}_\Gamma - L_{\Gamma\Gamma}\underline{\hat{\nu}}_\Gamma, \tag{39}$$

$$\underline{\widehat{V}}_\Gamma = N_{\Gamma\Gamma}\underline{\hat{\delta}}_\Gamma - M^t_{\Gamma\Gamma}\underline{\hat{\nu}}_\Gamma. \tag{40}$$

4.2 LSEMA module, test problem and results

The LSEMA subroutine computes the solution of Laplace's equation surrounding thin shells or discontinuities. As with the LBEMA, the subroutine relies on L3ALC to compute the matrix components in the systems (38)–(40). In this subsection, the LSEMA routine is demonstrated through solving a test problem.

The module LSEMA has the form:

```
LSEMA(MAXNODES,NNODE,NODES,MAXPANELS,NPANEL,PANELS,
  * MAXPOINTS,NPOINT,POINTS,
  * HA,HB,HF,HAA,HBB,HFF,
  * HIPHI,HIVEL,PINPHI,
  * LSOL,LVALID,TOLGEOM,
  * PHIDIF,PHIAV,VELDIF,VELAV,PPHI,
  * AMAT,BMAT,L_EH, M_EH,
  * PERM,XORY,C,WKSPC1,WKSPC2,WKSPC3).
```

The LSEMA parameters are similar to the LBEMA ones. However the expressions of the boundary condition and the boundary function are different.

HA stores the values of α on the shell panels, similarly HB, β; HAA, A; and HBB, B. The main output from the subroutine is PHIDIF that corresponds to $\underline{\hat{\delta}}_{\Gamma}$; PHIAV, $\underline{\hat{\Phi}}_{\Gamma}$; VELDIF, $\underline{\hat{v}}_{\Gamma}$; VELAV, $\underline{\hat{V}}_{\Gamma}$; and PPHI, $\underline{\hat{\varphi}}_{E}$.

The test problem is in file LSEMA_T. It consists of two circular coaxial parallel plates in the r, θ plane, of radius 1.0 and a distance of 0.1 apart in the planes where $z = 0.0$ and $z = 0.1$. A Dirichlet boundary condition is applied to both plates. On the plate at $z = 0.0$, the potential of 0.0 is applied, and a potential ($\delta = 0$, $\Phi = 0$) of 1.0 is applied on the other plate ($\delta = 0$, $\Phi = 1$). A complete analytic solution is not available. However in the central region between the plates, a simple gradient of potential is intuitive, as discussed. The results from the test problem are listed in **Table 6**.

4.3 LSEM3 module, test problem and results

The LSEM3 module solves Laplace's equation exterior to a thin shell in three dimensions. The subroutine call has the following form:

```
LSEM3(MAXNODES,NNODE,NODES,MAXPANELS,NPANEL,PANELS,
  * MAXPOINTS,NPOINT,POINTS,
  * HA,HB,HF,HAA,HBB,HFF,
  * HINPHI,HINVEL,PINPHI,
  * LSOL,LVALID,TOLGEOM,
  * PHIDIF,PHIAV,VELDIF,VELAV,PPHI,
  * AMAT,BMAT,L_EH,M_EH,
  * PERM,XORY,C,WKSPC1,WKSPC2,WKSPC3)
```

The definition of the important parameters can be found from the previous notes on LBEM3 and LSEMA. The test problem is in the file LSEM3_T, and it is similar to the test problem for LSEMA. This time, the open boundaries are two unit square plates of in $x - y$ planes. The two squares are 0.1 apart: one is at a potential of zero and the other is at a potential of one. The squares are each divided into 32 panels. The results at points between the squares, along a central axis, are shown in **Table 7**.

Index	Point	Expected (4 d.p.)	Computed (4 d.p.)
1	(0.0, 0.025)	0.25	0.2495
2	(0.0, 0.05)	0.5	0.5000
3	(0.0, 0.075)	0.75	0.7506

Table 6.
The results from the axisymmetric shell problem.

Index	Point	Expected (4 d.p.)	Computed (4 d.p.)
1	(0.5, 0.5, 0.1)	0.1	0.0962
2	(0.5, 0.5, 0.3)	0.3	0.02994
3	(0.5, 0.5, 0.5)	0.5	0.5000
4	(0.5, 0.5, 0.7)	0.7	0.7006
5	(0.5, 0.5, 0.9)	0.9	0.9041

Table 7.
The results from the three-dimensional shell problem.

5. Conclusions

In this paper a design of a software library has been set out and implemented in Fortran. In taking a 'library' approach, components can be developed that can be shared. There is, therefore, an overall reduction in coding, in line with good software engineering practice. For the three-dimensional problems, it is shown how exterior problems can be solved with the same code as interior problems. It is also shown how the core discrete operator components can be reused for codes solving problems in the same dimensional space. The method for solving the linear system of equations can also often be shared, as with LU factorisation, applied in these codes. A test problem has been developed in order to demonstrate each code. The library of codes and the way they are linked are set out in Appendix.

There are several areas for further development. It is good for software engineering also to widen participation to provide strong validation in the BEM, so that errors, for example, in the boundary mesh are noted before executing the BEM. In this work the validation is developed through the VGEOM* modules.

In this paper, the BEM codes have been applied to a set of simple test problems. It would be useful if a standard library of test problems emerged, so that all existing and future codes can be benchmarked against the same tests, with information such as error and processing time. More complex geometries—such as multiple surfaces in exterior problems or cavities in the domain for interior problems—would benefit from standard test problems. The codes are also adaptable to problems in which there is an existing field that the boundary and boundary conditions modify (via the *INPHI and *INVEL parameters), but these have not been tested.

Central to the efficiency of the method, as the number of elements increases, is the method for solving the linear system of equations and the method of storing the matrices. Computing the matrices in the BEM takes $O(n^2)$ time and memory. Solving the linear system by a direct method, like LU factorisation used in this work, takes $O(n^3)$ time. Hence, in order to scale up the method, LU factorisation needs to be replaced by an interative method, and methods of storing and computing the matrices may also become an issue.

In the software engineering approach in this work, a generalised form of the boundary condition is also operational, and interior and exterior problems in 3D are dealt with in the same code. Further generality may be achieved by forming a hybrid of the method that allows both open and closed surfaces [28–30].

A. Appendix

The main codes for solving Laplace problems by the boundary element method in this work are LIBEM2 for the two-dimensional problem interior to a closed

File/code	Purpose of module	LIBEM2	LBEMA	LBEM3	LSEMA	LSEM3
L2LC	Computes the discrete Laplace operators (2D)	X				
L3ALC	Computes the discrete Laplace operators (axisym)		X		X	
L3LC	Computes the discrete Laplace operators (3D)			X		X
GLS2	Solves a generalised linear system of equations	X	X	X	X	X
LUFAC	Carries out LU factorisation of the matrix	X	X	X	X	X
LUFBSUB	Carries out forward and back substitution	X	X	X	X	X
GEOM2D	Geometrical operations (2D)	X	X		X	
GEOM3D	Geometrical operations (3D)		X	X	X	X
GLRULES	Gauss-Legendre quadrature rules	X	X		X	
GLT7	7-point Gaussian quadrature rule for triangle			X		X
GLT25	25-point Gaussian quadrature rule for triangle			X		X
VGEOM2	Verifies the geometry (2D)	X				
VGEOMA	Verifies the geometry (axisym)		X		X	
VGEOM3	Verifies the geometry (3D)			X		X
VG2LC	Verifies the use of the L2LC module	X				
L3ALCC	Copy of L3ALC (to fake recursion)		X		X	

Table 8.
The main codes and supporting library.

boundary, LBEM3 for the general three-dimensional problem interior or exterior to a closed boundary, LBEMA for the axisymmetric three-dimensional problem interior or exterior to a closed boundary, LSEM3 for the general three-dimensional problem exterior to an open boundary and LSEMA for the axisymmetric three-dimensional problem exterior to an open boundary. The linkage between these and the supporting codes in the library is shown in **Table 8**.

The main subroutines have the control parameters LSOL, LVALID and TOLGEOM. LSOL is set to TRUE if the full solution is sought and FALSE if the linear system is the output. LVALID is set to TRUE if validation is required and FALSE if it is not. TOLGEOM sets the geometrical tolerance.

The GLS algorithm in file GLS2 carries out a column-swapping method [31] in order to prepare the linear system for solution by a standard method. The standard method in this work is LU factorization and back substitution in files LUFAC and LUFBSUBS.

Author details

Stephen Kirkup* and Javad Yazdani
University of Central Lancashire, Preston, UK

*Address all correspondence to: smkirkup@uclan.ac.uk

IntechOpen

© 2020 The Author(s). Licensee IntechOpen. This chapter is distributed under the terms of the Creative Commons Attribution License (http://creativecommons.org/licenses/by/3.0), which permits unrestricted use, distribution, and reproduction in any medium, provided the original work is properly cited.

References

[1] Wrobel LC, Aliabadi MH. The Boundary Element Method, Applications in Thermo-Fluids and Acoustics. Chichester, UK: John Wiley & Sons; 2002

[2] Li ZC. Combined Methods for Elliptic Equations with Singularities, Interfaces and Infinities. Vol. 444. Boston, London: Kluwer Academic Publishers; 1998

[3] Lindgren LE. From Weighted Residual Methods to Finite Element Methods. Available from: https://www.ltu.se/cms_fs/1.47275!/mwr_galerkin_fem.pdf [Accessed: 10 April 2019]

[4] Gato C. Detonation-driven fracture in thin shell structures: Numerical studies. Applied Mathematical Modelling. 2010;**34**:3741-3753

[5] Semwogerere T. Application analysis of the curved elements to a two-dimensional Laplace problem using the boundary element method. International Journal of Applied Physics and Mathematics. 2015;**5**:167-176

[6] Seydou F, Seppanen T, Ramahi O. Numerical solution of 3D laplace and helmholtz equations for parallel scatterers. In: Proceedings of the Antennas and Propagation Society International Symposium; Washington, DC, USA: IEEE; 2005. pp. 6-9

[7] Shahbazi M, Mansourzadeh S, Pishevar AR. Hydrodynamic analysis of autonomous underwater vehicle (AUV) flow through boundary element method and computing added-mass coefficients. International Journal of Artificial Intelligence and Mechatronics. 2015;**3**:212-217

[8] Weber C. Electron Energy Loss Spectroscopy with Plasmonic Nanoparticles. Graz, Austria: Karl-Franzens-Universität Graz; 2011

[9] Kirkup SM. DC capacitor simulation by the boundary element method. Communications in Numerical Methods in Engineering. 2007;**23**:855-869

[10] Cunderlik R, Mikula K, Mikula K. A comparison of the variational solution to the Neumann geodetic boundary value problem with the geopotential model. Contributions to Geophysics and Geodesy. 2004;**34**:209-225

[11] Sklyar O, Kueng A, Kranz C, Mizaikoff B, Lugstein A, Bertagnolli E, et al. Numerical simulation of scanning electrochemical microscopy experiments with frame-shaped integrated atomic force microscopy: SECM probes using the boundary element method. Analytical Chemistry. 2005;**77**:764-771

[12] Nkurunziza D, Kakuba G, Mango JM, Rugeihyamu SE, Muyinda N. Boundary element method of modelling steady state groundwater flow. Applied Mathematical Sciences. 2014;**8**:8051-8078

[13] Hashemi AR, Pishevar AR, Valipouri A, Părău EI. Numerical and experimental study on the steady cone-jet mode of electro-centrifugal spinning. Physics of Fluids. 2018;**30**:114108

[14] Kirkup SM, Henwood DJ. An empirical error analysis of the boundary element method applied to Laplace's equation. Applied Mathematical Modelling. 1994;**18**:32-38

[15] Hu H. An analysis of double Laplace equations on a concave domain. An Anal Double Laplace Equations a Concave Domain. 2018;**6**:304-314

[16] Menin OH, Rolnik V. Relation between accuracy and computational time for boundary element method applied to Laplace equation. Journal of

Computational Interdisciplinary Sciences. 2013;**4**:1-6

[17] Fiala P, Rucz P. NiHu: An open source C++ BEM library. Advances in Engineering Software. 2014;**75**:101-112

[18] Wieleba P, Sikora J. Open source BEM library. Advances in Engineering Software. 2009;**40**:564-569

[19] Kumara PK. A MATLAB code for three dimensional linear elastostatics using constant boundary elements. International Journal of Advanced Engineering. 2012;**2**:9-20

[20] Kirkup SM, Yazdani J. A gentle introduction to the boundary element method in Matlab/freemat. In: Proceedings of the WSEAS MAMECTIS; Corfu, Greece; 2008

[21] Kirkup SM. Fortran codes for computing the discrete Helmholtz integral operators. Advances in Computational Mathematics. 1998;**9**: 391-404

[22] Kirkup SM. The Boundary Element Method in Acoustics. Hebden Bridge: Integrated Sound Software; 1998

[23] Kirkup SM. The boundary element method in excel for teaching vector calculus and simulation. World Academy of Science, Engineering and Technology, International Science Index 144, International Journal of Social, Behavioral, Educational, Economic, Business and Industrial Engineering. 2019;**12**:1605-1613

[24] Kirkup SM. The boundary element method in acoustics: A survey. Applied Sciences. 2019;**9**:1642-1690

[25] Kirkup SM. Boundary Element Method. Available from: www.boundary-element-method.com [Accessed: 11 January 2018]

[26] Shaukat K, Tahir F, Iqbal U, Amjad S. A comparative study of numerical analysis packages. International Journal of Computer Theory and Engineering. 2019;**10**:67-72

[27] Warham AGP. The Helmholtz Integral Equation for a Thin Shell. London, UK: National Physical Laboratory; 1988

[28] Kirkup SM. The boundary and shell element method. Applied Mathematical Modelling. 1994;**18**:418-422

[29] Kirkup SM. The computational modelling of acoustic shields by the boundary and shell element method. Computers and Structures. 1991;**40**: 1177-1183

[30] Kirkup SM. Solution of discontinuous interior Helmholtz problems by the boundary and shell element method. Computer Methods in Applied Mechanics and Engineering. 1997;**140**:393-404

[31] Kirkup SM. Solving the linear systems of equations in the generalized direct boundary element method. In: Proceedings of the 1st International Conference on Numerical Modelling in Engineering; Ghent, Belgium; 2018

Chapter 2

Optimal Control of Thermal Pollution Emitted by Power Plants

Lèye Babacar, Tine Léon Matar and Sy Mamadou

Abstract

The coastal areas near thermal or nuclear plants are subject to hot water discharges produced by cooling processes. These activities induce an increase of the temperature near the outlet vicinity, which can extend for miles. The temperature variation affects the metabolic rate of organisms and the level of dissolved oxygen. Cooling by cold water from an additional discharge can be considered in order to limit this thermal pollution. This paper present a methodology based on the implementation of a two-dimensional numerical model to study the dynamic of the temperature originated from the industrial discharges. Moreover the optimal injection rate of cold water is sought to keep the water temperature as close as possible to the survival of the ecosystem. Numerical simulations are performed to illustrate the efficiency approach.

Keywords: CFD, power plant, modeling, thermal pollution, optimal control

1. Introduction

Thermal pollution is defined as the degradation of water quality by any process that changes ambient water temperature [1]. Coastal areas are often subject to thermal effluents originating from the cooling processes in industrial plants (nuclear reactors, electric power plants, petroleum refineries, steel melting factories, etc.) [2]. The industries collect water from lakes, rivers, or ocean, for cooling purpose, and return it in the environment at a high temperature. The hot water affects aquatic life, causes the substitution of aquatic fauna and flora, increases the mortality of certain species, and has indirect effects including bacterial development. More precisely, increasing the water temperature often increases the susceptibility of organisms to toxic substances (which are undoubtedly present in contaminated water) [3–6].

Studying the thermal effluents in receiving environments can contribute to efficiently manage the discharges, reducing environmental and economic impacts. Hence the reduction of thermal pollution must be included in the installation of cooling systems. Moreover distance between inlet and outlet must be carefully determined to avoid a decrease of the power plant efficiency.

By the middle of the 1960s, there were many research projects concentrating on thermal discharges, where major publications focus on the environmental impacts of power plant thermal discharges. Early mathematical models took place with works of [7]. The first treatments addressed the equilibrium iso-contour of elevated temperature within the receiving waters. Slightly later more advanced models allow the analysis of thermal plumes across extensive data in relation to seasonal and climate change fluctuations [3, 8–11].

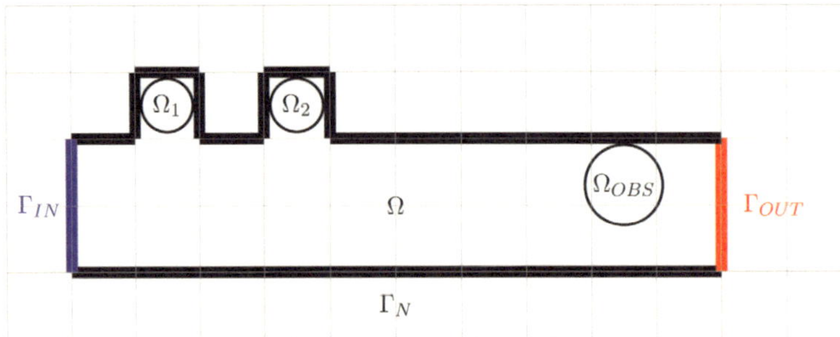

Figure 1.
Water domain Ω, industrial plants Ω_1, control zone Ω_2, observation zone Ω_{OBS}, impermeable boundary Γ_N, inflow boundary Γ_{IN}, and outflow boundary Γ_{OUT}.

In this research, thermal pollution due to industrial activities was modeled by a system of partial differential equations, and optimal control is applied to reduce the associated thermal pollution. The location of the understudy area is illustrated by **Figure 1**.

The paper is organized as follows. First, the thermal pollution is modeled by a coupling of Navier-Stokes and heat equations. The cold water injection rate is minimum of a cost function, in order to reduce the temperature variation and the energy required to refresh injected water. Afterward, the well-posedness of this problem is investigated. It follows a numerical resolution of the optimal control by means of a gradient descent algorithm. Finally, numerical simulations are performed to illustrate our approach.

2. CFD modeling of the thermal dispersion

2.1 Geometry representation

We are interested in the evolution of the system in space and time. Then, we denote x and t, respectively, as the space and time variables. Ω is the domain occupied by the water. Its boundary is denoted as $\partial\Omega$ and is divided into three disjoint subborders. It is written

$$\partial\Omega = \Gamma_N \cup \Gamma_{IN} \cup \Gamma_{OUT}, \tag{1}$$

where Γ_{IN} is the entering border, Γ_{OUT} is the outflow boundary, and Γ_N is the impermeable part. Ω contains three subdomains Ω_1, Ω_2, and Ω_{OBS}. Ω_1 stands for the industrial plants, where the source of pollution modeled by $f(x,t)$ is defined. In Ω_2, cold water is injected at a rate U in order to control the temperature in Ω_{OBS}. The objective consists in finding the optimal rate U_{opt} so that the temperature in Ω_{OBS} will be as close as possible to a desired value T_d. T_d can be the temperature favorable to the survival of the ecosystem. The geometric domain is illustrated by **Figure 1**.

2.2 Mathematical model

We present the system of partial differential equations representing the evolution of the river parameters (temperature, velocity, and pressure). Then, the cost function to be minimized in order to reduce the thermal pollution is described.

2.2.1 Temperature

By hypothesis, three processes influence the temperature evolution: the thermal conduction, the convection, and the internal reactions. The thermal conduction translates the fact that the heat flux is proportional to the temperature gradient. The convection expresses the temperature transfer by the fluid velocity. The internal reactions are represented by the different sources of temperature and the industrial plant discharges in this situation. By taking into account these processes, for a time $T_f > 0$, the temperature dynamic in $\Omega \times]0, T_f[$ is described by the equation

$$\underbrace{\frac{\partial T}{\partial t}}_{\text{variation}} - \underbrace{k\Delta T}_{\text{diffusion}} + \underbrace{u.\nabla T}_{\text{convection}} = \underbrace{f\psi_1}_{\text{discharges}} + \underbrace{U\psi_2}_{\text{control}}. \tag{2}$$

$T(x,t)$ represents the fluid temperature at position $x \in \Omega$ and time $t \in]0, T_f[$. k stands for the thermal diffusion coefficient. $u(x,t)$ is the fluid velocity inducing the advection process. The velocity is obtained by solving the Navier-Stokes system, described below. $\psi_1(x)$ and $\psi_2(x)$ are, respectively, Ω_1 and Ω_2 characteristic functions. They allow to localize the source term $f(t)$ and control $U(t)$, respectively, in the subdomains Ω_1 and Ω_2. The source term $f(t)$ is given, while the control $U(t)$ must be computed as a solution of an optimal control problem, described in the sequel. $T_{in}(x,t)$ is the temperature distribution in the inlet border:

$$T = T_{in} \quad \text{on} \quad \Gamma_{IN} \times]0, T[. \tag{3}$$

On the impermeable boundary, no heat flux boundary condition is considered:

$$-k\frac{\partial T}{\partial \vec{n}} = 0 \quad \text{on} \quad \Gamma_N \times]0, T[, \tag{4}$$

where the vector $\vec{n}$ defined on the boundary constitutes the outward unit normal vector. On the outflux boundary, the heat flux is proportional to the velocity and the temperature:

$$-k\frac{\partial T}{\partial \vec{n}} + \alpha_T\left(u.\,\vec{n}\right)T = 0 \quad \text{on} \quad \Gamma_{OUT} \times]0, T[, \tag{5}$$

where $\alpha_T > 0$ is a constant. The boundary condition allows us, as we will see in the sequel in Subsection 3.2, to obtain an explicit formula for the cost function gradient. These boundary conditions for the temperature are summarized in **Figure 2**.

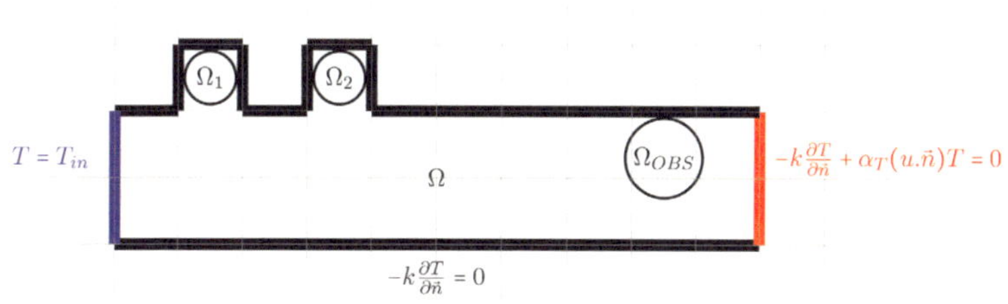

Figure 2.
Boundary conditions for temperature.

Function $T_0(x)$ represents the distribution of the temperature at the initial time:

$$T(0) = T_0 \text{ on } \overline{\Omega}. \tag{6}$$

2.2.2 Velocity and pressure

The fluid velocity u is obtained by solving, in $\Omega\times]0, T_f[$, the incompressible Navier-Stokes system:

$$\begin{aligned} \frac{\partial u}{\partial t} - \nu\Delta u + (u.\nabla)u + \nabla p &= f_1\psi_1 + f_2\psi_2, \\ div(u) &= 0, \end{aligned} \tag{7}$$

where $p(x,t)$ is the water pressure; $\nu > 0$ is the kinematic viscosity; $f_1(t)$ and $f_2(t)$ are, respectively, the velocity sources in Ω_1 and Ω_2. At the inlet, the velocity is known and given by a function $u_{in}(x,t)$. It is written

$$u = u_{in} \text{ on } \Gamma_{IN}\times]0, T_f[. \tag{8}$$

On the impermeable boundary, it is assumed that the velocity is equal to zeros due to the viscosity:

$$u = 0 \text{ on } \Gamma_N\times]0, T_f[. \tag{9}$$

On the outflow boundary, the pressure is equal to zeros:

$$p = 0 \text{ on } \Gamma_{OUT}\times]0, T_f[, \tag{10}$$

The boundary conditions applied to the velocity and the pressure are summarized in **Figure 3**.

The system is also completed by the initial condition for the velocity:

$$u(0) = u_0 \text{ on } \overline{\Omega}. \tag{11}$$

2.2.3 Cost functional

In order to reduce the pollution in an arbitrary area Ω_{OBS}, a freshwater is introduced in the subdomain Ω_2. We are seeking the optimal rate U at which the freshwater is introduced, such that the temperature in Ω_{OBS} must be as closed as possible to a prescribed threshold denoted T_d. This optimal control must be the minimum of the cost function:

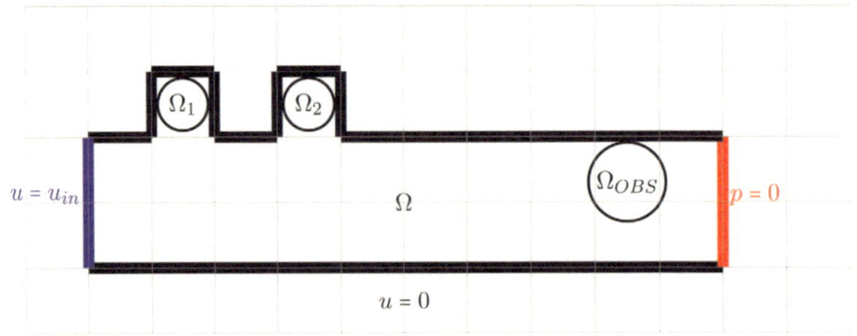

Figure 3.
Boundary conditions for the velocity and pressure.

$$J(T,U) = \frac{1}{2}\left(\int_0^{T_f}\int_{\Omega_{OBS}} (T - T_d)^2 dxdt + \beta \int_0^{T_f}\int_{\Omega_2} (U - U_d)^2 dxdt\right), \tag{12}$$

where U_d is the ideal control rate and $\beta > 0$ is the cost-efficiency ratio. More β is great; more energy must be provided to maintain the temperature in Ω_{OBS} close to T_d.

3. Optimal control

3.1 Cost functional

The aim is to find an optimal control U minimizing the cost function:

$$\tilde{J}(v) = J(T(v), v), \quad \forall v \in \mathcal{U}_{ad}, \tag{13}$$

where $\mathcal{U}_{ad} = L^2(\Omega_2)$ is the admissible function space. By considering the symmetric, continuous, coercive bilinear form

$$\pi(s,v) = \; < T(s) - T_0, T(v) - T_0 > + \beta < s - U_d, v - U_d > \tag{14}$$

for all $s, v \in U_{ad}$, and the linear bounded functional

$$F(v) = < T_d - T_0, T(v) - T_0 >, \tag{15}$$

the cost function is written

$$\tilde{J}(v) = \frac{1}{2}\left(\pi(v,v) + \|T_d - T_0\|^2\right) - F(v). \tag{16}$$

We are in the framework of Theorem 16.1 in [12] that establishes the existence and uniqueness of solution to the minimization problem.

3.2 Directional derivative

First, for a fixed $h \in L^2(]0, T_f[, \mathcal{U}_{ad})$, two function sequences

$$T_\lambda = T(U + \lambda h), \quad w_\lambda = T_\lambda - T \tag{17}$$

are considered, for all $\lambda > 0$.

3.2.1 Sequences convergence

The difference between equations satisfied by T_λ and T results to the following one:

$$\begin{aligned} \frac{\partial w_\lambda}{\partial t} - k\Delta w_\lambda + u.\nabla w_\lambda &= \lambda h \psi_2 \text{ in } \Omega \times]0, T_f[, \\ w_\lambda &= 0 \text{ on } \Gamma_{IN} \times]0, T_f[, \\ -k\frac{\partial w_\lambda}{\partial \vec{n}} &= 0 \text{ on } \Gamma_N \times]0, T_f[, \end{aligned} \tag{18}$$

$$-k\frac{\partial w_\lambda}{\partial \vec{n}} + \alpha_T\left(u.\,\vec{n}\right)w_\lambda = 0 \text{ on } \Gamma_{OUT}\times]0,T_f[,$$
$$w_\lambda(0) = 0 \text{ on } \overline{\Omega}. \tag{19}$$

If $u \in L^\infty(\Omega\times]0,T_f[)$, this previous system admits a unique weak solution satisfying

$$\|w_\lambda\|_{L^2(]0,T_f[,V)} + \|w_\lambda\|_{C([0,T_f],H)} + \|w_\lambda\|_{L^2(]0,T_f[,L^2(\Gamma_{OUT}))} \leq C\lambda\|h\|_{L^2(]0,T_f[,\mathcal{U}_{ad})}. \tag{20}$$

with $C > 0$ [13]. The functional spaces are defined by $H = L^2(\Omega)$ and

$$V = \left\{v \in H^1(\Omega) \text{ such that } v = 0 \text{ on } \Gamma_{IN}\right\}. \tag{21}$$

It can be deduced from the preceding inequality that

$$\lim_{\lambda\to 0} w_\lambda = 0 \text{ in } L^2(]0,T_f[,V) \cap C([0,T_f],H). \tag{22}$$

3.2.2 Directional derivative computation

A direct computation gives us

$$\tilde{J}(U+\lambda h) - \tilde{J}(U) = \frac{1}{2}\int_0^{T_f} (<(T_\lambda + T - 2T_d)\psi_{OBS}, w_\lambda> + <\lambda\beta(2(U-U_d)+\lambda h)\psi_2, h>)dt. \tag{23}$$

By dividing this last equality by λ, it becomes

$$\frac{\tilde{J}(U+\lambda h) - \tilde{J}(U)}{\lambda} = \frac{1}{2}\int_0^{T_f} (<(T_\lambda + T - 2T_d)\psi_{OBS}, w_u> + <\beta(2(U-U_d)+\lambda h)\psi_2, h>)dt, \tag{24}$$

where $w_u = \frac{w_\lambda}{\lambda}$ is solution of the equation:

$$\frac{\partial w_u}{\partial t} - k\Delta w_u + u.\nabla w_u = h\psi_2 \text{ in } \Omega\times]0,T_f[,$$
$$w_u = 0 \text{ on } \Gamma_{IN}\times]0,T_f[,$$
$$-k\frac{\partial w_u}{\partial \vec{n}} = 0 \text{ on } \Gamma_N\times]0,T_f[, \tag{25}$$
$$-k\frac{\partial w_u}{\partial \vec{n}} + \alpha_T\left(u.\,\vec{n}\right)w_u = 0 \text{ on } \Gamma_{OUT}\times]0,T_f[,$$
$$w_u(0) = 0 \text{ on } \overline{\Omega}.$$

By passing to the limit $\lambda \to 0$, the directional derivative is written

$$\tilde{J}'(U)\cdot h = \int_0^{T_f} (<(T - T_d)\psi_{OBS}, w_u> + <\beta(U-U_d)\psi_2, h>)dt. \tag{26}$$

Unfortunately, this directional derivative does not provide an explicit expression of the gradient. To achieve this, the term

$$\int_0^{T_f} < (T - T_d)\psi_{OBS}, w_u > dt \tag{27}$$

must be written as a scalar product on h. In this scope, the Lagrangian approach is used and consists of solving an adjoint system, stated below.

3.3 Explicit gradient

Equation (25) is multiplied by the adjoint function $\tilde{p}$ and integrated over $\Omega\times]0, T_f[$. The result is

$$\int_0^{T_f} < h\psi_{\Omega_2}, \tilde{p} > dt = \int_0^{T_f} < \frac{\partial w_u}{\partial t} - k\Delta w_u + u.\nabla w_u, \tilde{p} > dt. \tag{28}$$

Integrations by parts lead to

$$\int_0^{T_f} < h\psi_{\Omega_2}, \tilde{p} > dt = \int_0^{T_f} < -\partial\frac{\tilde{p}}{\partial t} - k\Delta\tilde{p} - u.\nabla\tilde{p}, w_u > dt + I_1 + I_2, \tag{29}$$

where

- $I_1 = \int_\Omega (w_u(T_f)\tilde{p}(T_f) - w_u(0)\tilde{p}(0))dx,$
- $I_2 = \int_0^{T_f} \int_{\partial\Omega} \left(\left(k\partial\frac{\tilde{p}}{\partial\vec{n}} + \alpha_T \left(u.\ \vec{n}\right)\tilde{p}\right) w_u - k\frac{\partial w_u}{\partial\vec{n}}\tilde{p} \right) d\sigma dt.$

By using the initial and boundary conditions of w_u, the terms I_1 and I_2 become

- $I_1 = \int_\Omega w_u(T_f)\tilde{p}(T_f)dx,$
- $I_2 = \int_0^{T_f} \left(-\int_{\Gamma_{IN}} k\frac{\partial w_u}{\partial\vec{n}}\tilde{p}d\sigma + \int_{\Gamma_N} \left(k\frac{\partial\tilde{p}}{\partial\vec{n}} + \alpha_T\left(u.\ \vec{n}\right)\tilde{p}\right) w_u d\sigma + \int_{\Gamma_{OUT}} k\frac{\partial\tilde{p}}{\partial\vec{n}} w_u d\sigma \right) dt.$

From the condition $u = 0$ on Γ_N, it becomes

$$I_2 = \int_0^{T_f} \left(-\int_{\Gamma_{IN}} k\frac{\partial w_u}{\partial\vec{n}}\tilde{p}d\sigma + \int_{\Gamma_N \cup \Gamma_{OUT}} k\frac{\partial\tilde{p}}{\partial\vec{n}} w_u d\sigma \right) dt.$$

Hence, we assume that $\tilde{p}$ is solution of the adjoint problem:

$$\begin{aligned} -\frac{\partial\tilde{p}}{\partial t} - k\Delta\tilde{p} - u.\nabla\tilde{p} &= (T - T_d)\psi_{OBS} \text{ in } \Omega\times]0, T_f[, \\ \tilde{p} &= 0 \text{ on } \Gamma_{IN}\times]0, T_f[, \\ k\frac{\partial\tilde{p}}{\partial\vec{n}} &= 0 \text{ on } (\Gamma_N \cup \Gamma_{OUT})\times]0, T_f[, \\ \tilde{p}(T) &= 0 \text{ on } \overline{\Omega}. \end{aligned} \tag{30}$$

Consequently, it becomes $I_1 = 0$, $I_2 = 0$, and

$$\int_0^{T_f} < h\psi_{\Omega_2}, \tilde{p} > dt = \int_0^{T_f} < (T - T_d)\psi_{OBS}, w_u > dt. \tag{31}$$

Using this above equality in relation (26), we obtain

$$\tilde{J}'(U) \cdot h = \int_0^{T_f} < (\tilde{p} + \beta(U - U_d))\psi_{\Omega_2}, h > dt, \tag{32}$$

hence

$$\nabla \tilde{J}(U) = (\tilde{p} + \beta(U - U_d))\psi_{\Omega_2}. \tag{33}$$

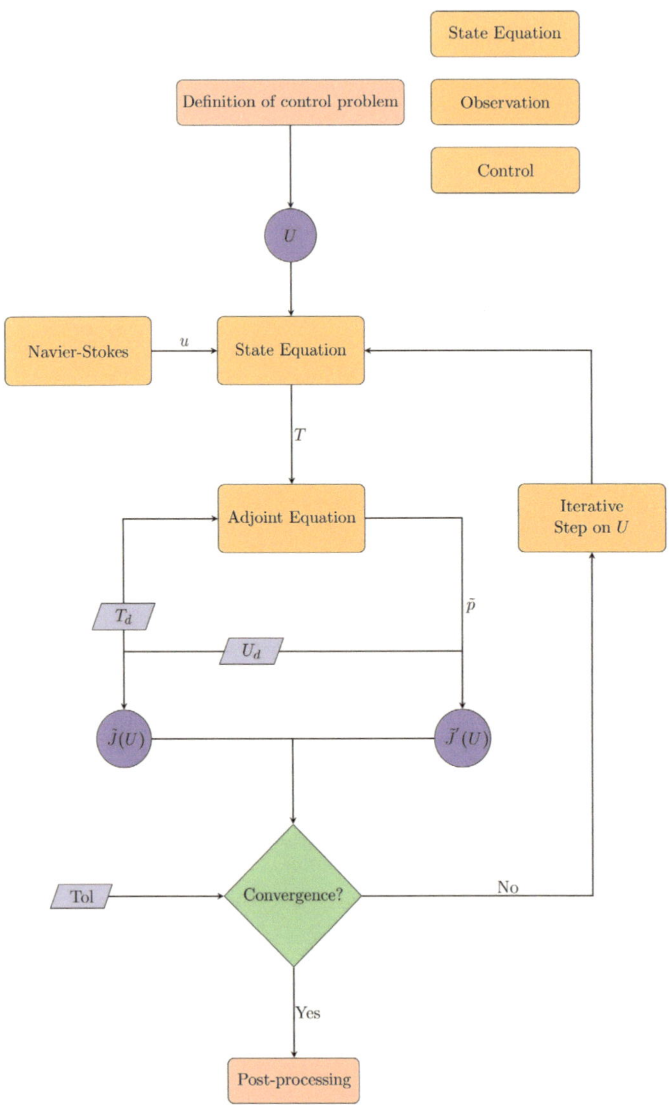

Figure 4.
Flow chart of the iterative algorithm of the solution to the optimal control problem.

A change of variables $\overline{p}(x,t) = \tilde{p}\left(x, T_f - t\right)$ is made where $\overline{p}$ is the solution of

$$\begin{aligned}\left(\frac{\partial \overline{p}}{\partial t} - k\Delta\overline{p} - u.\nabla\overline{p}\right)(t) &= \psi_{OBS}(T - T_d)\left(T_f - t\right) \text{ in } \Omega\times]0, T_f[,\\ \overline{p} &= 0 \text{ on } \Gamma_{IN}\times]0, T_f[,\\ k\frac{\partial \overline{p}}{\partial \vec{n}} &= 0 \text{ on } (\Gamma_N \cup \Gamma_{OUT})\times]0, T_f[,\\ \overline{p}(0) &= 0 \text{ on } \overline{\Omega}.\end{aligned} \tag{34}$$

The gradient becomes

$$\nabla\tilde{J}(U)(t) = \left(\overline{p}\left(T_f - t\right) + \beta(U - U_d)(t)\right)\psi_{\Omega_2}. \tag{35}$$

This gradient allows to solve the minimization problem (12). The gradient descent algorithm is used to compute the optimal control.

3.4 Iterative algorithm

First, the Navier-Stokes system is solved on $\overline{\Omega} \times \left[0, T_f\right]$ to obtain the fluid velocity. Secondly, the optimal control $U(t), t \in \left[0, T_f\right]$, is computed by means of a descent algorithm with a fixed step. And finally, this optimal control is used in the state equation to simulate the fluid temperature propagation. The optimal control is the limit of the sequence:

$$U^0(t) \in \mathcal{U}_{ad}, \quad U^{m+1}(t) = U^m(t) - \tau\nabla\tilde{J}(U^m)(t), \tag{36}$$

τ being the step. The algorithm used is described as follows:
Input: Initial control: $U^0(t)$, Maximal number of iterations: $m_{\max}$, Tolerance: *tol*.
This algorithm is summarized by **Figure 4**.

$m \leftarrow 0.$

while $m < m_{max}$ *and* $\frac{|\nabla\tilde{J}(U^m)(t)|}{|\nabla\tilde{J}(U^0)(t)|} > tol$ **do**

  Solve the state equation

$$\frac{\partial T}{\partial t} - k\Delta T + u.\nabla T = f\psi_1 + U^m\psi_2,$$

  with initial and boundary conditions.

  Solve the adjoint problem

$$\left(\frac{\partial \bar{p}}{\partial t} - k\Delta\bar{p} - u.\nabla\bar{p}\right)(t) = \psi_{OBS}(T - T_d)(T_f - t),$$

  with initial and boundary conditions.

  Compute the new control

$$U^{m+1}(t) = U^m(t) - \tau\nabla\tilde{J}(U^m)(t).$$

  $m \leftarrow m + 1.$

end

Algorithm 1.
Optimal control algorithm.

4. Numerical scheme

4.1 State equation

The state equation is solved by using a method of $\mathbb{P}_1$ discontinuous Galerkin in space and implicit finite difference in time. The fluid velocity is very high in relation to its thermal conductivity. To stabilize the induced oscillations, streamline diffusion [12] is introduced in the scheme; hence the solved state equation is as follows:

$$\frac{\partial T}{\partial t} - k\Delta T + u.\nabla T + \underbrace{\frac{H}{|u|} u\Delta(uT)}_{\text{streamline diffusion}} = f\psi_1 + U\psi_2, \tag{37}$$

H being the maximal mesh element diameter.

4.2 Adjoint equation

As the state equation, the adjoint problem is solved by using a method of $\mathbb{P}_1$ discontinuous Galerkin in space and implicit finite difference in time. Streamline diffusion is introduced in the scheme; hence the solved adjoint state is as follows:

$$\left(\frac{\partial \overline{p}}{\partial t} - k\Delta \overline{p} - u.\nabla \overline{p} + \underbrace{\frac{H}{|u|} u\Delta(u\overline{p})}_{\text{streamline diffusion}} \right)(t) = \psi_{OBS}(T - T_d)\left(T_f - t\right). \tag{38}$$

4.3 Navier-Stokes system

The Navier-Stokes system is solved by means of a P_1 Lagrange finite element method for the velocity and the pressure. The following algorithm proposed by Chorin [14] is used for the time discretization:

1. Computation of an intermediate solution u^*

$$\frac{u^* - u^n}{\Delta t} = -(u^n.\nabla)u^n + \nu\Delta u^n + \tilde{f}n,$$

2. Computation of the pressure p^{n+1}

$$\Delta p^{n+1} = \frac{1}{\Delta t}\nabla.u^*,$$

3. Computation of the velocity u^{n+1}

$$u^{n+1} = u^* - \Delta t \nabla p^{n+1},$$

where u^n, p^n, and $\tilde{f}n$, $n \in \mathbb{N}^*$, are, respectively, the approximated velocity, pressure, and source term at the nth time step. The mesh is frequently adapted to improve the solution efficiency. For the numerical implementation, the solver of partial differential equations FreeFem++ downloadable at http://www.freefem.org/ff++

allows to do the domain meshing, the computation, and the post-processing of the solution. The numerical code is run on a computer of characteristics ProBook 250 G2, processor Intel(R) Core(TM) I5-5200U CPU @ 2.20 GHz 2.20 GHz, and RAM memory 8.00 Go.

5. Numerical results

This section presents numerical tests to illustrate the validity of our approach. The river parameters are listed in **Table 1**.

The initial temperature is always constant and equal to $T_0 = 30°C$. The initial velocity is given by $u_0 = 0 m.s^{-1}$. The temperature at the inlet boundary is set to $T_{in} = 30°C$, while the velocity profile is described by the parabolic function:

$$h(\mathbf{x}, \mathbf{y}, t) = u_{\max} * \mathbf{y} * (2 - \mathbf{y}) \quad m.s^{-1}. \tag{39}$$

$u_{\max}$ is the maximal value of the velocity. At the outflow boundary, mixed boundary conditions are used with $\alpha_T = 10^{-8}$ and $\alpha_u = 10^{-8}$. The velocity source at the discharge is given by

$$f_1 = (0, -v_{\max}) \quad m.s^{-1}, \tag{40}$$

with $v_{\max} > 0$. For the optimal control, the target temperature in the observation area is equal to $T_d = 30°C$, and the target control is of $U_d = 0°C.s^{-1}$. The cost-efficiency ratio of the objective functional is defined by $\beta = 1$. The time step is set to $\Delta t = 0.1\,s$. The stopping criteria tolerance of the iterative algorithm is given by $tol = 0.02$, and the step of the descent gradient algorithm by $\tau = 0.5$.

5.1 Thermal pollution simulation

We assume that hot water is discharged in Ω_1 by power plants. The distribution of the water temperature at different time steps is shown in **Figure 5**. The flow velocity is presented in **Figure 6**. It can be observed that flow displaces the thermal plume from the power plants to the right hand side.

5.2 Influence of the discharge rates

In **Figure** 7, we present the influence of different discharge rates on the thermal plume area. For low rates, we observe a high temperature far from the discharge area. However, for high rates, the temperature seems to be high near the emission zone and small far from the discharge zone. In this last case, the polluted area is more extended.

Parameters	Notation	Value	Unit
Viscosity	ν	$8.84 * 10^{-4}$	$m^2.s^{-1}$
Thermal diffusion	k	$1.5 * 10^{-7}$	$m^2.s^{-1}$

Table 1
Physical parameters of the river.

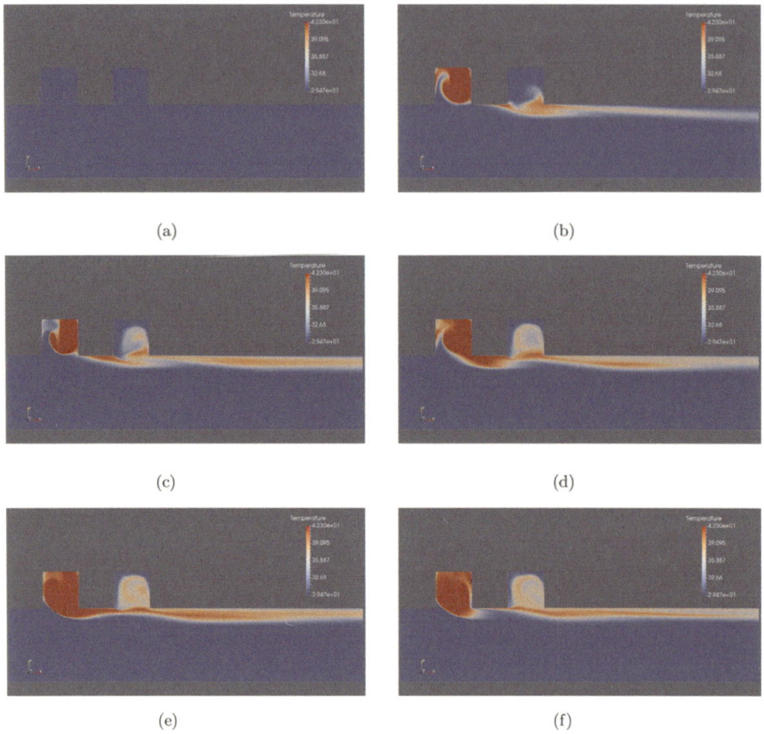

Figure 5.
Temperature contours. (a) Temperature contour after 0 second, (b) Temperature contour after 10 second, (c) Temperature contour after 20 second, (d) Temperature contour after 30 second, (e) Temperature contour after 40 second, and (f) Temperature contour after 50 second.

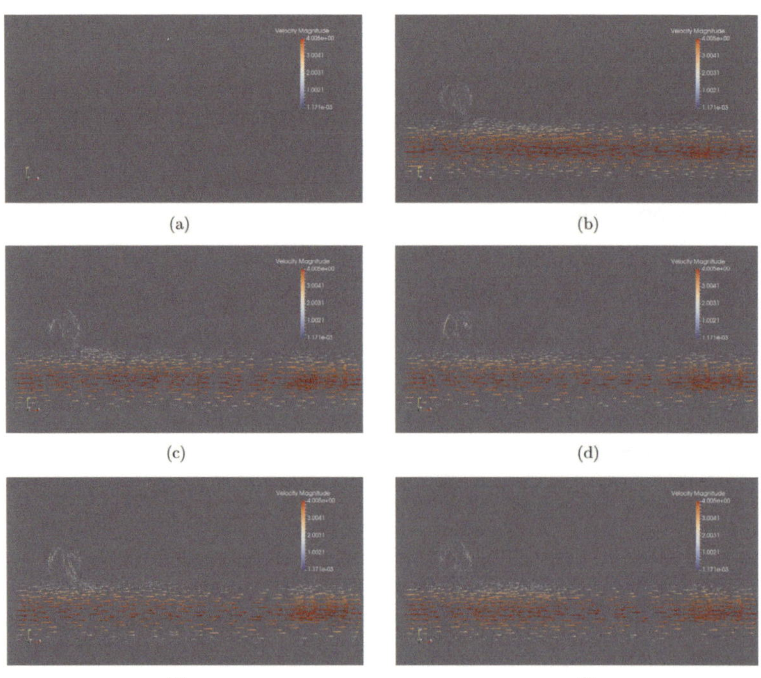

Figure 6.
Velocity fields. (a) Velocity field after 0 second, (b) Velocity field after 10 second, (c) Velocity field after 20 second, (d) Velocity field after 30 second, (e) Velocity field after 40 second, and (f) Velocity field after 50 second.

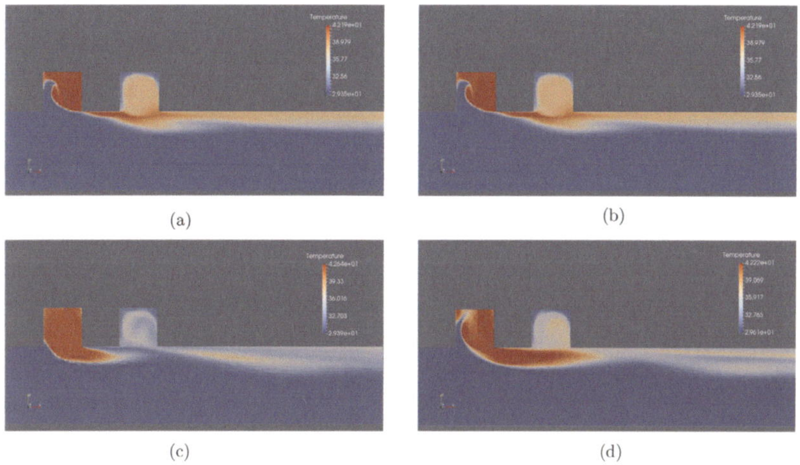

Figure 7.
Temperature contours for different discharge rates. (a) $v_{max} = 20$, *(b)* $v_{max} = 30$, *(c)* $v_{max} = 40$, *and (d)* $v_{max} = 50$.

5.3 Heat emission optimal control

Similarly to the first case, a source term f is applied on Ω_1. The control on Ω_2 is initialized to $U = -1°C.s^{-1}$. The velocity sources on Ω_1 and Ω_2 are, respectively:

$$f_1 = (0, -10)m.s^{-1}, \quad f_2 = (0, -10)m.s^{-1}. \tag{41}$$

The cost function value according to the optimization iteration is represented by **Figure 8**. At the initial step, we assume that $U = -1°C.s^{-1}$, thus obtaining a cost function $J(T, U) = 1.37274 * 10^{-2}$ (**Figure 9**). The optimal solution is obtained after 10 iterations, for an optimal control rate $U(t)$ of order -10^{-2} illustrated by **Figure 12**.

In **Figure 10**, the stopping criteria $|\nabla\tilde{J}(U^m)|/|\nabla\tilde{J}(U^0)|$ in terms of the number of iterations are reported. At initial step, the stopping criteria is equal to 1 and then decreases to reach $9.56428 * 10^{-3}$ after 10 iterations. From this observation, it can be deduced that the control sequence U^m converges to the optimal control rate.

Figure 11 compares the temperature evolution in Ω_{OBS} for the thermal plume dispersion simulation $U = 0°C.s^{-1}$, the initial step $U = -1°C.s^{-1}$, and the optimal

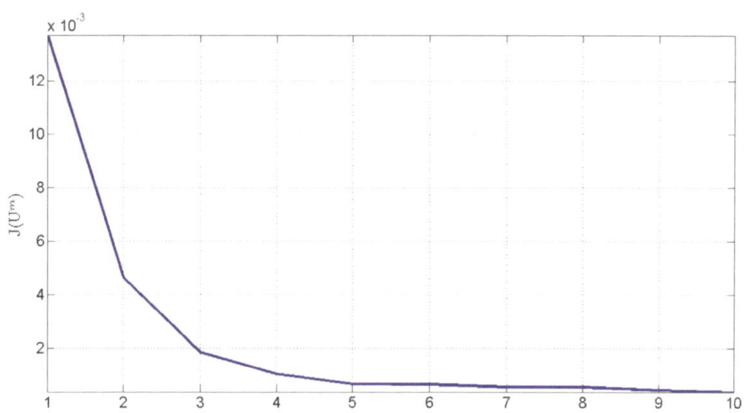

Figure 8.
Cost functional after each iteration of the steepest descent algorithm.

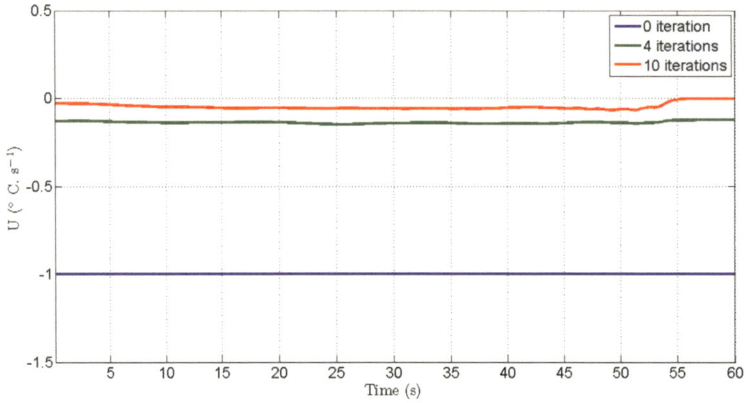

Figure 9.
Control rate after iterations 0, 4, and 10.

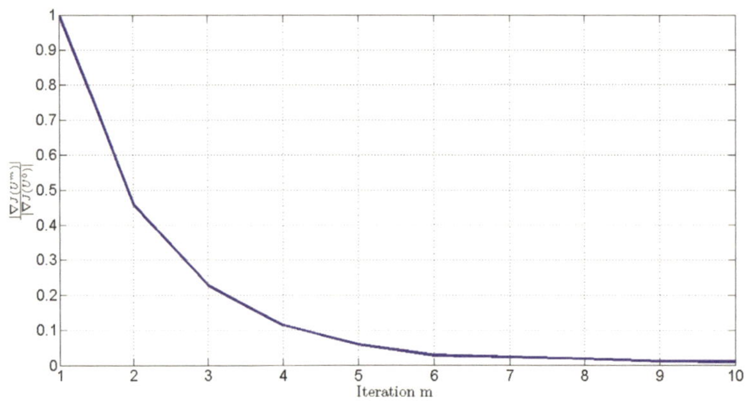

Figure 10.
Stopping criteria according to the number of iterations.

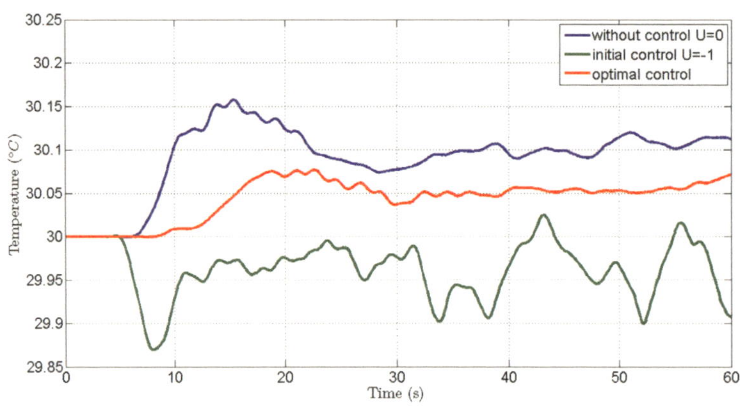

Figure 11.
Time series of water mean temperature in the observation zone Ω_{OBS}. In blue, only hot water discharge in Ω_1 is considered. In green, simulations are carried out by using the initial control $U = -1°C.s^{-1}$. In red, temperature evolution is corresponding to the optimal control.

case. For the thermal dispersion, it can be noticed that after 8 s the temperature increases to reach a maximum of 30.15°C. For the initial step $U = -1°C.s^{-1}$, the temperature is lower than 30°C and reaches a minimum of 29.87. For the optimal

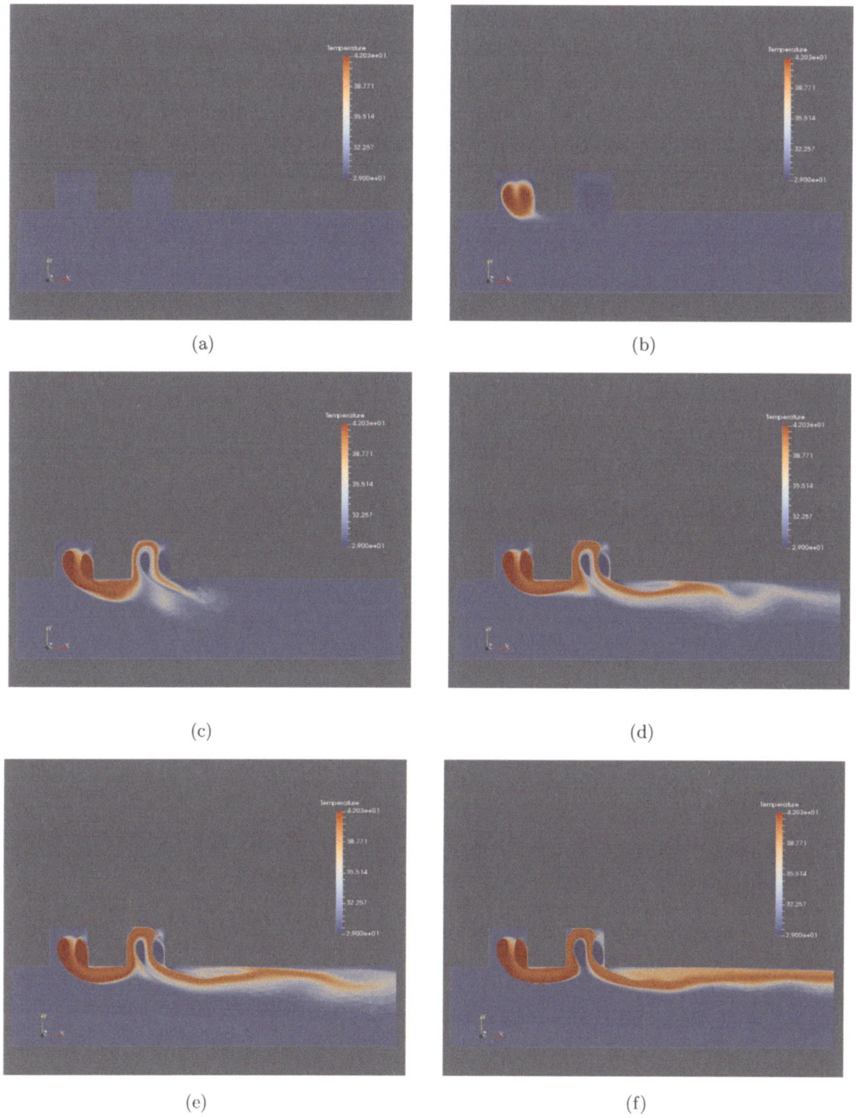

Figure 12.
Optimal control: temperature field (in °C) at time intervals of 1 (a), 5 (b), 10 (c), 30 (d), 45 (e), and 59.9 s (f).

solution, the temperature is $>30°C$ but does not exceed $30.08°C$. Moreover for all the time, it is closer to the optimal value than for the thermal dispersion case. According to these remarks, it can be concluded that the computed optimal rate allows to maintain the temperature in Ω_{OBS} at a value close to the desired threshold $30°C$.

Figure 12 illustrates the temperature distributions at times 1, 5, 10, 30, 45, and 59.9 s. A reduction of the thermal pollution is observed, due to the cold water source in Ω_2.

6. Conclusion

Numerical models are essential to predict the thermal effluent impacts on natural systems. This work is of particular relevance for the coastal area managements, by contributing to a better understanding of these consequences on coastal dynamics. The cooling water discharge causes an increase of the water temperature. Model simulations show that water dynamic plays a significant role on the temperature dispersion. The optimal control of the model allows to define a strategy to limit this pollution. The simulations show that an injection of freshwater, at an appropriate rate, allows to reduce this pollution and keeps water temperature favorable for ecosystem survival.

Author details

Lèye Babacar[1*], Tine Léon Matar[2] and Sy Mamadou[3]

1 Laboratoire Eaux, Hydro-Système et Agriculture, Rue de la Science, Institut International d'Ingénierie de l'Eau et de l'Environnement, Ouagadougou, Burkina Faso

2 Institut Camille Jordan UMR 5208, Université Claude Bernard Lyon 1, Villeurbanne, France

3 Laboratoire d'Analyse Numérique et d'Informatique, Université Gaston Berger, Saint-Louis, Sénégal

*Address all correspondence to: babacar.leye@2ie-edu.org

IntechOpen

© 2019 The Author(s). Licensee IntechOpen. This chapter is distributed under the terms of the Creative Commons Attribution License (http://creativecommons.org/licenses/by/3.0), which permits unrestricted use, distribution, and reproduction in any medium, provided the original work is properly cited.

References

[1] Precht H, Christophersen J, Hensel H, Larcher W. Heat Exchange with the Environment. Berlin, Heidelberg: Springer; 1973. pp. 545-564

[2] Issakhov A, Zhandaulet Y. Numerical study of technogenic thermal pollution zones' formations in the water environment from the activities of the power plant. Environmental Modeling and Assessment. 2019;**24**:1-16

[3] Abbaspour M, Javid AH, Moghimi P, Kayhan K. Modeling of thermal pollution in coastal area and its economical and environmental assessment. International journal of Environmental Science and Technology. 2005;**2**(1):13-26

[4] Coulter DP, Sepúlveda MS, Troy CD, Höök TO. Thermal habitat quality of aquatic organisms near power plant discharges: Potential exacerbating effects of climate warming. Fisheries Management and Ecology. 2014;**21**(3): 196-210

[5] Hester ET, Doyle MW. Human impacts to river temperature and their effects on biological processes: A quantitative synthesis 1. JAWRA Journal of the American Water Resources Association. 2011;**47**(3):571-587

[6] Kelso JRM, Milburn GS. Entrainment and impingement of fish by power plants in the great lakes which use the once-through cooling process. Journal of Great Lakes Research. 1979;**5**(2):182-194

[7] Edinger JE, Brady DK, Geyer JC. Heat Exchange and Transport in the Environment. Report No. 14. Technical Report. NTIS Issue No. 198314. 1974, p. 137

[8] El-Ghorab EAS. Physical model to investigate the effect of the thermal discharge on the mixing zone (case study: North Giza Power Plant, Egypt). Alexandria Engineering Journal. 2013; **52**(2):175-185

[9] Hunt CD, Mansfield AD, Mickelson MJ, Albro CS, Geyer WR, Roberts PJW. Plume tracking and dilution of effluent from the Boston sewage outfall. Marine Environmental Research. 2010;**70**(2):150-161

[10] Kalinowska MB, Rowiński PM. Thermal Pollution in Rivers—Modelling of the Spread of Thermal Plumes. Cham: Springer International Publishing; 2015. pp. 591-613

[11] Issakhov A, Zhandaulet Y. Numerical simulation of thermal pollution zones' formations in the water environment from the activities of the power plant. Engineering Applications of Computational Fluid Mechanics. 2019;**13**(01):279-299

[12] Quarteroni A. Numerical Models for Differential Problems. New York, USA: Springer; 1988

[13] Brézis H. Analyse fonctionelle: Collection Mathématiques appliquées pour la matrise. New York, USA: Masson; 1983

[14] Chorin A. Numerical solution of the Navier-Stokes equations. Mathematics of Computation. 1968;**22**:10

Chapter 3

Finite Difference Solution of Conjugate Heat Transfer in Double Pipe with Trapezoidal Fins

Ghazala Ashraf, Khalid S. Syed and Muhammad Ishaq

Abstract

A conjugate heat transfer problem on the shell side of a finned double pipe heat exchanger is numerically studied by suing finite difference technique. Laminar flow with isothermal boundary conditions is considered in the finned annulus with fully developed flow region to investigate the influence of variations in the fin height, the number of fins and the fluid and wall thermal conductivities on the hydraulic and thermal design of the exchanger. The governing momentum and energy equations have been solved by a finite difference-based numerical algorithm. The improvement in heat transfer rates, the exchanger performance and the optimum configurations are discussed.

Keywords: incompressible, laminar flow, convective heat transfer, finned double pipe, fully developed flow, conjugate heat transfer, isotherms

1. Introduction

The theory of heat transfer is widely used in many fields of engineering industries and also in applied sciences. Heat exchangers are being used in power generation houses and nuclear reactor centres, in order to generate and convert energy for unlimited purposes. The design of the heat exchangers, according to its usage, also is a matter of great importance.

As for boiling, condensing and radiating the fluid and other things, the size of heat transfer equipment is always taken into account. In aerospace, equipment requires the limitation of weights, while in nuclear reactors, a deep study of heat transfer analysis is needed to avoid the unbearable damages [1]. In heating, generally these heat exchangers show the very low rate of heat transfer. The performance of such heat exchanger can be signified by various techniques. Özisik [2] has given a detailed study of augmented fin surfaces which are of great help in enhancement of heat transfer rate. A similar study of heat flow was made by Nasiruddin and Kamran [3], for vortex generation by applying baffles in circular ducts. A study on convective heat transfer with variable fin heights was made by Zeitoun and Hegazy [4], in which a rise in heat transfer rate was observed, with low friction factor. Suryanarayana and Apparao [5] mentioned in his work that one of the criteria for evaluating the performance of a heat exchanger with extended surfaces is the pumping power required for a specified heat duty. He reported that average heat transfer coefficient increases with an increase in the frequency of the number of fins. Another study for heat transfer was made by using elliptic pin finned tube by Qingling et al. [6]. Adegun et al. [7] proposed a new method to increase the heat transfer rate by using circular pipes making them inclined, with different inclinations. They investigated that heat transfer rate is rapid till 15° inclination and fin height H = 0.2 and increase in fin height is just a waste of material and causes more expenses. But before the last few decades, it was not found by any good mathematical approach. Pagliarini [8] indicated that the idealization of infinite conduction may cause unrealistic approach in the analysis of heat transfer characteristics. So, they proposed the idea of finite conductivity offered by the material used in it.

The transfer of heat between fluid and solid while flowing in any heat exchanger is governed by two different kinds of equations, as transfer of heat in fluid is governed by the elliptical Navier-Stokes equation or by the parabolic boundary layer equation, and the heat transfer inside the body is governed by the elliptical Laplace equation or by the parabolic differential equation [9]. This forms a so-called model of conjugate problem. Conjugate heat transfer problems have been analysed in various geometric configurations. Kumar [10] examined two conjugate problems of heat transfer in the laminar boundary layer at the boundary of a semi-infinite porous medium on the assumption that fluid filters continuously through the porous surface and that the injection velocity varies as $x^{-1/2}$. Barozzi and Pagliarini [11] used finite element method to examine a conjugate problem of a laminar flow in a pipe when the outer wall is being heated uniformly to observe the effects of wall conduction.

Mori et al. [12] investigated the conjugate problem in a circular pipe and observed the conduction effects. In this study, it was proposed that conductivities of the wall and of fluid make remarkable affects in heat transfer properties when thickness of the wall is increased, while thin wall effects on the heat transfer properties are negligible. An analysis on conjugate heat transfer by using three types of boundary conditions, as constant heat flux, constant wall temperature and constant heat transfer coefficients, was made by Sakakibara et al. [13]. They used Duhamel's theorem to calculate the interfacial temperature. They reported that conduction in the wall is inversely proportional to the ratio of conductivities of solid to fluid.

Kettner et al. [14] numerically investigated that the ratio of thermal conductivities of the solid to fluid has no noticeable effect when the fins of the small height are considered. However, this conductivity ratio has a significant effect when the fin height relative to pipe radius is taken more than 0.4. A similar conjugate problem was studied in finned tube, and it was reported that fin efficiency has a great influence on heat flux and heat transfer coefficient by Fiebig et al. [15]. The conjugate heat transfer problem was investigated in different geometries by Nguyen et al. [16]. Nordstorm and Berg [17] investigated the Navier-Stokes equations for modified interface conditions. They have computed conjugate problem with two approaches: one is by using heat equation for the transfer of heat in solid, and the other is transfer of heat by using Navier-Stokes equations.

Sohail and Fakhir [18] gave numerical investigation of double-pipe heat exchanger with circumferential fins in longitudinal to study the effect of fin pitch-to-height (P/H) ratio on heat transfer and fluid flow characteristics at various Reynolds numbers, using water as the working fluid. Systematic analysis is carried out by changing geometric and flow parameters. Geometric parameters include varying the pitch-to-height ratio from 0.55 to 26.4, while for the flow parameters, Reynolds number varied from 200 to 1400.

Syed et al. [19] made numerical simulation of finned double-pipe heat exchanger, where fins are distributed around the outer wall of the inner pipe. By using H1 (constant heat flux) and T1 (constant wall temperature) boundary conditions and one-dimensional fin equation, he concluded that the fin heat loss increases if we increase the conductivity of fin. Mazhar [20] extended the work of Syed [19] by considering laminar conjugate heat transfer in the thermal entrance region of the finned annulus. In his work, finite difference method (FDM) was used for the simulation of the problem of hydrodynamically fully developed and thermally developing and fully developed flow. The investigation made by him was clearly showing that the entrance region is more affected by the ratio of conductivities and rate of heat transfer rate at entrance is higher than that of the fully developed region.

The methods in the previous discussion are conventional methods for enhancing heat transfer rate. They have their own limitations such as the need of extended surfaces, enhancement in thermal processing equipment sizes and increase in pumping power to acquire the desirable efficiency level.

2. Iterative methods

In this chapter, a numerical algorithm for solving elliptic PDEs involving fluid flow and of heat transfer analysis is described. This algorithm uses multigrid discretization for nested iterations to accelerate the rate of convergence at higher levels with less computation. The well-known successive over-relaxation (SOR) method is used to solve the problem, by giving a fixed value to the relaxation parameter. Problems of steady-state viscous flow and steady temperature can be brought into the category of elliptic PDEs with appropriate boundary conditions. In order to solve these PDEs numerically, a higher order accuracy rate with less computation is more preferred. Moreover, estimation of error helps to ensure the accuracy of the solution.

Iterative methods are widely used in order to solve the difference equations, which are obtained from elliptic PDEs. Among these iterative methods, SOR method is widely used for its fast convergence for a class of large linear systems arising from difference equations. SOR method is a quick solver for a large number of linear equations.

The momentum and energy equations are solved by using the algorithm given in the next section, for the behaviour of fully developed laminar flow through a finned double pipe. A comparative study of literature results and the present work are shown in "Results" section.

2.1 Iterative scheme for Poisson equation

A general Poisson equation in two dimensions $\nabla^2 u = f(x,y)$ (1)

can be approximated by the pictorial relation $\nabla^2 u = \frac{1}{h^2}\left\{\begin{matrix} & 1 & \\ 1 & -4 & 1 \\ & 1. & \end{matrix}\right\}$

The function $u(x,y)$ can be replaced by the value at the discrete nodes of the region. In order to discretize the function, a square grid with step size h can be helpful. The value of the exact solution function $u(x,y)$ at a point $P(x_i, y_j)$ is denoted by $u\left(x_i, y_j\right)$, and its approximated value is denoted as $u(i,j)$. For the chosen discretization of the function, the partial differential equations are approximated at the grid points by using the discrete value of the function $u(i,j)$. The first- and second-order partial derivatives are approximated by the difference quotients. For this purpose central difference quotients are used as follows:

$$u_x\left(x_i, y_j\right) = \frac{u_{i+1,j} - u_{i-1,j}}{2h} \tag{2}$$

$$u_y\left(x_i, y_j\right) = \frac{u_{i,j+1} - u_{i,j-1}}{2h} \tag{3}$$

$$u_{xx}\left(x_i, y_j\right) = \frac{u_{i+1,j} - 2u_{i,j} + u_{i-1,j}}{h^2} \tag{4}$$

$$u_{yy}\left(x_i, y_j\right) = \frac{u_{i,j+1} - 2u_{i,j} + u_{i,j-1}}{h^2} \tag{5}$$

The Poisson equation can be approximated by using the above central difference quotients:

$$\frac{u_{i+1,j} - 2u_{i,j} + u_{i-1,j}}{h^2} + \frac{u_{i,j+1} - 2u_{i,j} + u_{i,j-1}}{h^2} = \mathrm{f}\left(x_i, y_j\right) \tag{6}$$

The boundary conditions of the boundary value problem need to be applied in accordance with the difference approximation of PDEs. The simplest of the boundary conditions is the Dirichlet boundary condition. In this case the difference equation can be applied to all interior points with unknown functions, and known values of the function at each boundary can be directly replaced.

Here, a simple type of Neumann boundary condition is taken into consideration. It is considered that boundary is taken at grid parallel to any of the axes here, e.g., x-axis. This boundary condition requires the normal derivative to be disappeared. This derivative in normal direction can be approximated by the central difference quotients.

$$\text{So, we approximate } \left.\frac{\partial u}{\partial n}\right|_{\mathrm{P}} = 0 \text{ as } \frac{u_{i+1,j} - u_{i-1,j}}{2h} = 0 \tag{7}$$

$$\text{which implies } u_{i+1,j} = u_{i-1,j} \tag{8}$$

The vanishing of normal derivative means that function $u(x, y)$ is symmetric about the boundary of the region.

2.2 Linear system solver

SOR is an iterative method, which is an important solver for the class of large linear system arising from the finite difference approximation of PDEs. It is not only an efficient solver but also a smoother. The method of relaxation is an iterative scheme which permits one to select the best equation to be used for the faster rate of convergence. Although Gauss Jacobi and Gauss Seidel are taken as one of the good iterative methods, relaxation method is more advantageous because of its faster rate of convergence depending upon the relaxation parameter 'ω.' When $0 < \omega < 1$, the procedures are called under relaxation methods. When $\omega = 1$, then relaxation method is same as that of Gauss-Seidel method.

For $1 < \omega < 2$, the procedure is called over-relaxation.

The general Laplace equation is discretized by using finite differences, and the boundary conditions are approximated by second-order central differences. Fictitious points lying outside the domain, arising from the discretized form of derivative boundary conditions, are expressed in terms of the points and then incorporated into the governing finite difference equations. In order to have faster rate of convergence, the discretized form of Laplace equation in Eq. (9) can be approximated by using SOR method, which may take the form

$$u_{i,j}^{n+1} = u_{i,j}^{n} + \omega/4\left[\left(f\left(x_i, y_j\right)\right) - 4u_{i,j}^{n} - a_{i+1,j}u_{i+1,j}^{n} - a_{i-1,j}u_{i-1,j}^{n+1} - a_{i,j+1}u_{i,j+1}^{n} - a_{i,j-1}u_{i,j-1}^{n+1}\right] \tag{9}$$

where the value of relaxation factor 'ω' is obtained by hit and trial method in order to have faster rate of convergence.

The SOR iteration procedure can be terminated by using the following convergence criteria depending upon the convergence nature of the system being solved. In order to have computational results with a minimum number of iterations, absolute norm is used to decide the convergence criteria, which is given below:

$$e(i,j) = \left|\overline{u_{i,j}} - u_{i,j}\right| < \varepsilon \tag{10}$$

where ε is the order of convergence for each iteration and its value is taken as $\varepsilon = 0.00001$.

3. Problem formulation

In convective forced flow, the assumptions of negligible wall thickness and infinite conductivity of the fin and wall may cause unrealistic predictions of heat transfer characteristics, because of such ideal assumptions. In the present study, we take into account finite wall thickness of the inner pipe wall for realistic results. Also, we consider finite conductivity of the fin and wall in order to avoid the overestimates of heat transfer characteristics. This require coupling of the heat conduction problem in the wall-fin assembly and convective heat transfer problem in the fluid, and this coupled system is called conjugate heat transfer problem. The constant heat flux boundary condition is applied to the inner side of the inner pipe, and adiabatic thermal condition is applied at the inner wall of the outer pipe. At each solid-fluid interface, the heat flux is continuous. Moreover, temperature is assumed to be continuous.

In order to describe the conjugate problem mathematically, we need to describe the momentum, energy and heat equations that are to be solved simultaneously. The momentum and energy equations will be described in the prescribed sections and will be transformed into their dimensionless forms by the means of dimensionless variables. Similarly, heat equation is treated and solved for the transfer of heat in the solid part of the domain. A cross-sectional view of finned double pipe (FDP) is shown in **Figure 1**.

The velocity field is then independent of the temperature field. Geometrical symmetries shown in **Figure 1** permit the equations to be only solved in the region $a \le r \le b$ and $0 \le \theta \le \alpha + \beta$ shown in **Figure 2**.

3.1 Momentum, heat and energy equations

The governing momentum, heat and energy equations in dimensionless form are given in the Eqs. (11)–(13), as below by using the transformations defined in Syed [19].

This problem is constrained in the region where $r_i \le r \le r_o$ and $0 \le \theta \le \alpha + \beta$ because of its geometrical symmetry as shown in **Figure 2**:

$$\frac{\partial^2 u^*}{\partial R^2} + \frac{1}{R}\frac{\partial u^*}{\partial R} + \frac{1}{R^2}\frac{\partial^2 u^*}{\partial \theta^2} = \frac{4}{c} \tag{11}$$

where $C = -\left(1 - R_m^2 + 2R_m^2 \ln R_m\right)$

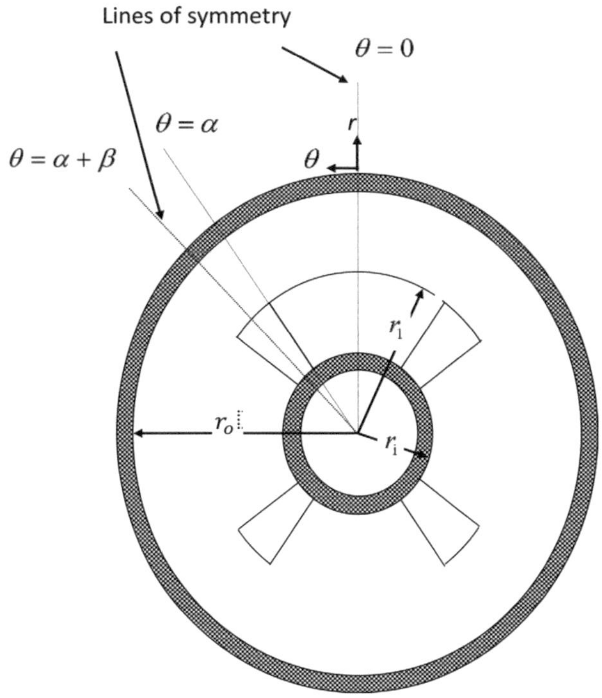

Figure 1.
Cross section of the finned double pipe.

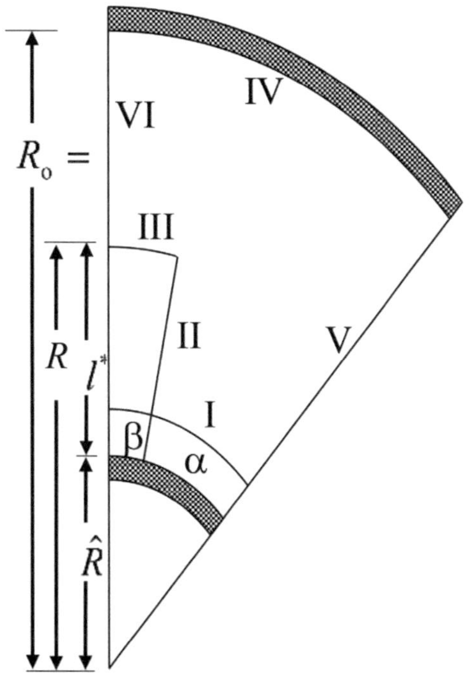

Figure 2.
Computational domain.

transformed into dimensionless form:

$$\frac{\partial^2 \tau^f}{\partial R^2} + \frac{1}{R}\frac{\partial \tau^f}{\partial R} + \frac{1}{R^2}\frac{\partial^2 \tau^f}{\partial \theta^2} = \frac{u^*}{A_c^* \overline{u}^*} \tag{12}$$

$$\frac{\partial^2 \tau^s}{\partial R^2} + \frac{1}{R}\frac{\partial \tau^s}{\partial R} + \frac{1}{R^2}\frac{\partial^2 \tau^s}{\partial \theta^2} = 0 \tag{13}$$

where $\overline{u}^*$ is the dimensionless mean velocity.

The boundary conditions are applied due to the viscosity of the fluid and symmetry offered by the domain shown in **Figure 2**. The boundary conditions in dimensionless form may be written as:

a. No slip conditions at the solid boundaries:

$$\text{(I) } u^* = 0 \text{ at } R = \hat{R},\ 0 \le \theta \le \alpha$$
$$\text{(II) } u^* = 0 \text{ at } \theta = \alpha,\ \hat{R} \le R \le R_1$$
$$\text{(III) } u^* = 0 \text{ at } R = R_1,\ \alpha \le \theta \le \alpha + \beta$$
$$\text{(IV) } u^* = 0 \text{ at } R = 1,\ 0 \le \theta \le \alpha + \beta$$

b. Symmetry conditions:

$$\text{(V) } \partial u^*/\partial \theta = 0 \text{ at } \theta = 0,\ \hat{R} \le R \le R_1$$
$$\text{(VI) } \partial u^*/\partial \theta = 0 \text{ at } \theta = \alpha + \beta,\ R_1 \le R \le 1$$

Constant flux boundary condition at the inner surface of the inner pipe

$$\tau = 0 \text{ at R} = R_w,\ 0 \le \theta \le \alpha \tag{14}$$

An adiabatic wall temperature condition at the inner surface of the outer pipe

$$\frac{\partial \tau}{\partial \theta} = 0 \text{ at R} = 1,\ 0 \le \theta \le \alpha + \beta \tag{15}$$

Boundary conditions at the lines of symmetry

$$\frac{\partial \tau^s}{\partial \theta} = 0 \text{ at } \theta = 0,\ R_w \le R \le R_o \tag{16}$$

$$\frac{\partial \tau^f}{\partial \theta} = 0 \text{ at } \theta = 0,\ \hat{R} \le R \le 1 \tag{17}$$

$$\frac{\partial \tau^s}{\partial \theta} = 0 \text{ at } \theta = \alpha + \beta,\ R_w \le R \le R_1 \tag{18}$$

$$\frac{\partial \tau^f}{\partial \theta} = 0 \text{ at } \theta = \alpha + \beta,\ \hat{R} \le R \le 1 \tag{19}$$

There are three interfaces where the solid and fluid mediums are contact. These three interfaces are termed here as inner pipe, fin lateral surface and fin tip interfaces. These may be defined as

$$\text{Inner pipe interface is at } r = r_i \& 0 \le \theta \le \alpha. \tag{20}$$

$$\text{Fin lateral surface interface is at } \theta = \alpha \& r_i \le r \le r_1 \tag{21}$$

$$\text{Fin tip interface is at } r = r_1 \text{ and } \alpha \le \theta \le \alpha + \beta. \tag{22}$$

On these interfaces, we impose the conditions of continuity of temperature and that of heat flux in order to maintain the energy balance. These interface conditions are used to couple conduction (Eq. (13)) in the solid with the energy (Eq. (12)) in the fluid. This forms the so-called conjugate problem.

The interface conditions can be expressed mathematically as given below by using same dimensionless transformations.

Continuity of fluxes at the solid-fluid interfaces

$$\frac{\partial \tau^s}{\partial R} = \frac{1}{\Omega}\frac{\partial \tau^f}{\partial R} \text{ at } R = \hat{R} \text{ and } 0 \le \theta \le \alpha \tag{23}$$

$$\frac{\partial \tau^s}{\partial R} = \frac{1}{\Omega}\frac{\partial \tau^f}{\partial R} \text{ at } R = R_1 \text{ and } \alpha \le \theta \le \alpha + \beta \tag{24}$$

$$\frac{\partial \tau^s}{\partial \theta} = \frac{1}{\Omega}\frac{\partial \tau^f}{\partial \theta} \text{ at } \theta = \alpha \text{ and } \hat{R} \le R \le R_1 \tag{25}$$

where $\Omega = \frac{\lambda^f}{\lambda^s}$ is the ratio of conductivities of fluid by solid.

Continuity of temperature at the solid-fluid interfaces

$$\tau^s = \tau^f \text{ at } R = \hat{R} \text{ and } 0 \le \theta \le \alpha \tag{26}$$

$$\tau^s = \tau^f \text{ at} R = R_1 \text{ and } \alpha \le \theta \le \alpha + \beta \tag{27}$$

$$T^s = T^f \text{ at } \theta = \alpha \text{ and } \hat{R} \le R \le R_1 \tag{28}$$

3.2 Numerical solutions

For numerical domain, Poisson equation given in Eq. (1) takes the form

$$\frac{U_{i+1,j} - 2U_{i,j} + U_{i-1,j}}{h^2} + \frac{1}{R_i}\frac{U_{i+1,j} - U_{i-1,j}}{2h} + \frac{1}{R_i^2}\frac{U_{i+1,j} - 2U_{i,j} + U_{i-1,j}}{k^2} = \frac{4}{C} \tag{29}$$

After combining the coefficient, we get

$$\left(\frac{1}{h^2} + \frac{1}{2R_ih}\right)U_{i+1,j} + \left(\frac{-2}{h^2} - \frac{2}{R_i^2k^2}\right)U_{i,j} + \left(\frac{1}{h^2} - \frac{1}{2R_ih}\right)U_{i-1,j} + \frac{1}{R_i^2k^2}U_{i,j+1} + \frac{1}{R_i^2k^2}U_{i,j-1} = \frac{4}{C} \tag{30}$$

For our simplicity, writing the coefficient in some standardized form

$$\mathrm{eU_{i+1,j} + pU_{i,j} + wU_{i-1,j} + nU_{i,j+1} + sU_{i,j-1} = rhs}$$

$$\text{where } e = \left(\frac{1}{h^2} + \frac{1}{2R_ih}\right), p = \left(\frac{-2}{h^2} - \frac{2}{R_i^2k^2}\right),$$

$$\mathrm{w} = \left(\frac{1}{\mathrm{h}^2} - \frac{1}{2\mathrm{R_ih}}\right), \mathrm{n} = \frac{1}{\mathrm{R_i^2k^2}}, \mathrm{s} = \frac{1}{\mathrm{R_i^2k^2}}, \mathrm{rhs} = \frac{4}{\mathrm{C}}$$

For energy and heat equations, similar scheme is developed.

3.3 Error analysis and validity of results

The numerical algorithm described in Chapter 2 has been used to determine the numerical results in the present study. The iterative convergence and interpolation procedure need to be validated. The error criterion given as $\varepsilon = 0.00001$ was used to terminate the iterative procedures. The present results of the friction factor and Nusselt number for copper have been compared with literature results for β = 3°. They were verified by comparing them with results present in literature with same geometrical parameters of heat exchanger. The present results differ with the literature results by less than 0.5%. However, in two exceptional cases, the results are not up to the same level of accuracy. The improved results obtained at all grid levels are of comparable accuracy. The comparison given in **Table 1** confirms the validity of numerical algorithm. This difference between friction factor and Nusselt number gives the overview; the percentage difference greater than '0' depicts the overestimation in present results, and difference less than '0' shows the underestimated values of the present study. Since the difference in the values provided in **Table 1** is negligible, this gives the validity of results in the present study.

4. Result and discussion

4.1 Local results

4.1.1 Flow behaviour

In this section, velocity contours are given with respect to different geometrical variations. **Figure 3a** and **b** show the velocity contours for $\hat{R} = 0.5, \beta = 2°, H^* = 0.6$. The effect of the number of fins is observed by firstly taking $M = 6$ and then $M = 18$.

While observing the contours, it is clear that between two consecutive fins a region of high velocity exists in the middle of annulus.

4.1.1.1 Velocity contours

Figure 3a shows the velocity contours, For M=6, the annulus region is filled with closed loops in the middle of the region, while near the inner wall of outer pipe circular loops are formed. For this fin height, two dimensional effects are more towards the outer pipe.

Figure 3b shows the velocity contours for $M = 18$, for $H^* = 0.6$. For the increased number of fins, the middle of region is surrounded by annular loops, which depict

Geometrical parameters		Comparison of friction factor			Comparison of Nusselt number		
M	H^*	*f*Re	*f*Re	Age change (%)	Nu(Old)	Nu(Cu)	Age change (%)
6	0.2	21.117	21.1351	0.0857	4.5189	4.5212	0.0508
	0.4	20.401	20.4185	0.0857	4.1846	4.1881	0.0836
	0.6	18.848	18.8592	0.0594	3.9827	3.9873	0.1154
	0.8	15.783	15.7868	0.0240	3.766	3.3789	−10.278
12	0.2	19.125	19.1580	0.1725	3.8433	3.4873	−9.262
	0.4	19.358	19.3991	0.2123	3.1644	3.1705	0.1928
	0.6	20.027	20.0653	0.1912	3.9256	3.9421	0.4203
	0.8	17.145	17.1592	0.0828	4.2311	4.2401	0.2127

Table 1.
Validity of present results of momentum equation and energy equation for $\beta = 3°$, $\hat{R} = 0.5$.

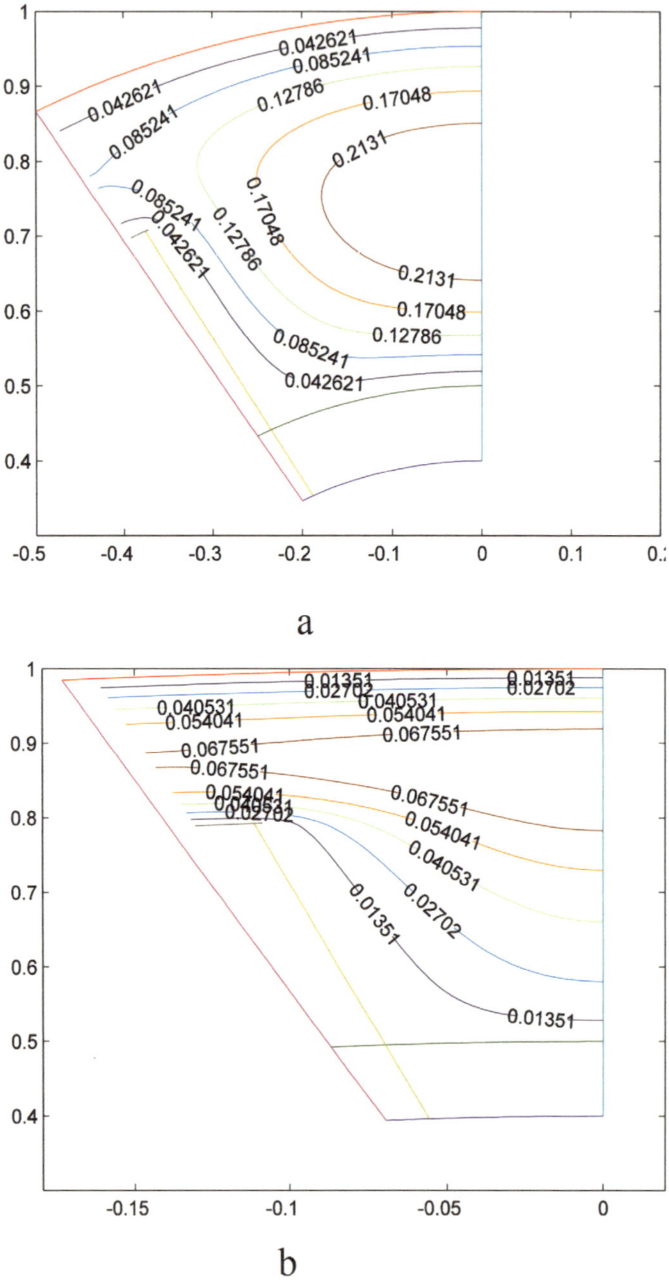

Figure 3.
(a) Velocity contours for $\hat{R}$ = 0.5, H^* = 0.8, M = 6. (b) Velocity contours for $\hat{R}$ = 0.5, H^* = 0.8, M = 6.

that the region of high velocity starts to develop at the middle of annulus. Also, when a number of fins are increased, then closed loop will break into annular loops.

We must have to notice that increase in the number of fins will also give rise to the value of friction factor, which may slow down the fluid motion in the pipe.

4.1.2 Heat transfer analysis

The results of energy and heat equations are combined together for analytical study of conjugate heat transfer, while fin and wall are offering finite heat conduction because of material used in them.

The same geometrical parameters and grid points are taken as that for momentum equation. A comparative study on behaviour of heat transfer is shown in **Figure 4a** and **b**. While using copper for fin and wall geometry, it should be noted that the isotherms corresponding to the zero level represent the geometry. Because of the definition of the dimensionless temperature used, the higher the isotherm level depicts, the lower the value of local temperature. The isotherms corresponding to value 1 have local fluid temperature equal to mean fluid temperature. Those with values lower than 1 have the local temperature greater than the mean temperature, and while greater than 1 indicate that the local temperature is lesser than the meant temperature. From these figures, it is clear that the region of high-temperature gradient near the inner pipe wall and near fin surface shows the high rate of convection. In order to comprise with interfaces and transfer of heat between fluid and solid, continuity of fluxes and of temperature is considered.

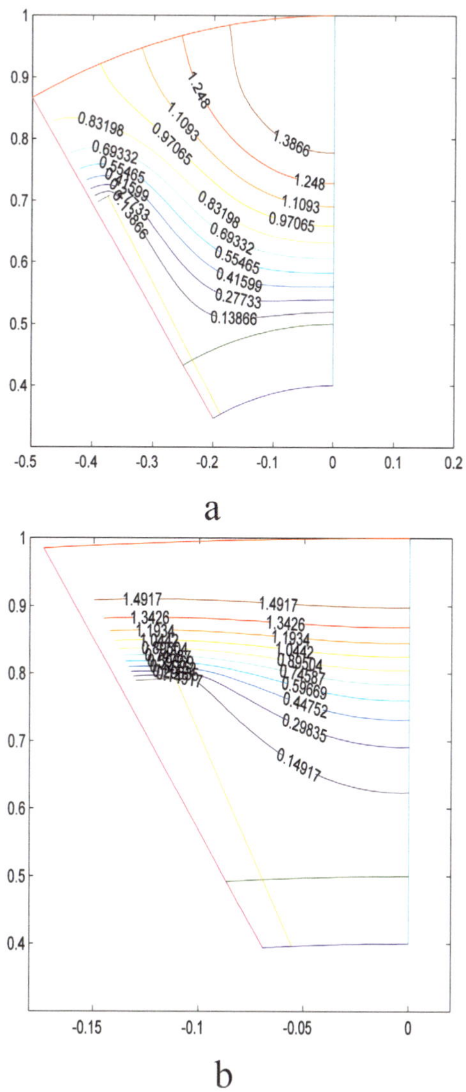

Figure 4.
(a) Isotherms (copper) for $\hat{R} = 0.5$, $H^* = 0.6$, $M = 6$. *(b) Isotherms (copper) for* $\hat{R} = 0.5$, $H^* = 0.6$, $M = 18$.

4.1.2.1 Isotherms

Figure 4a shows the isotherms for $M = 6$, $\hat{R} = 0.5$, $H^* = 0.6$, which shows that a region of higher temperature starts to develop in the middle of annulus and significant change occurs in temperature gradient. Whereas, **Figure 4b** shows the isotherms, when $M = 18$, the temperature gradient is higher in the entire cross section, and higher temperature region appears near the outer wall of the inner pipe, while lower near the inner wall of the outer pipe.

Apparently, in these isotherms conduction in wall and fin assembly is not being shown, but if we increase the number of contours, the temperature gradient is shown in the solid part of the domain too. This is because of the very high-temperature gradient in fluid and because of convection and conduction.

The trend of isotherms is more understandable corresponding to the velocity contours. From these figures, it is observed that the region of high-velocity gradients near the heated surfaces of the inner pipe and fin also has high-temperature gradients indicating the high rate of convection.

There is an equivalent transfer of heat in fin and fluid on the interfaces; interfaces get smoother because of the continuity of fluxes. But in case of temperature difference, it rises more rapidly in fluid than the solid (wall and fin).

4.2 Overall results

Figure 5a shows the plots of different values of Nusselt number of copper against the number of fins $H^* \in \{0.2, 0.4, 0.6, 0.8\}$ taking fixed value of ratio of radii $\hat{R} = 0.5$. For fin height $M = 6$, the value of Nusselt number decreases as we increase fin height. The plot of Nusselt number against these values shows monotonic decreasing behaviour. However, the value of Nu(Cu) remains higher for the $H^* = 0.2$ than the fin height $H^* = 0.4$ and others.

The plots of Nusselt numbers for $M = 6$ and $M = 18$ clearly depict that the behaviour of Nusselt number significantly differs in these two graphs. An optimal value of Nusselt number is observed for some particular value of $H^* = 0.2$ used. Thus, for $M = 6$, shorter fin gives better rate of heat conduction.

While taking the higher number of fins into account, for $M = 18$, Nusselt number gradually decreases when fin height is increased from $H^* = 0.2$ to $H^* = 0.4$. Its shows a parabolic behaviour. After reaching its lowest value, it starts increasing and attains its optimal value at the longest fin.

It can be concluded that the value of Nusselt number for $\hat{R} = 0.5$ at $H^* = 0.6$ and $H^* = 0.8$ is the best choice when $M = 9$ and $M = 18$ are taken, respectively.

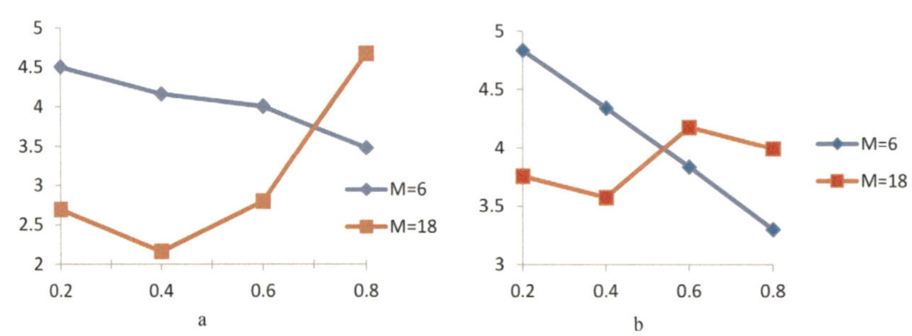

Figure 5.
(a) H^*-Nu(Cu), for $\hat{R} = 0.5$. (b) M-Nu(Cu), for $\hat{R} = 0.7$.

Figure 5b shows the plots of different values of Nusslet number of copper against number of fins $H^* \in \{0.2, 0.4, 0.6, 0.8\}$ taking fixed value of ratio of radii $\hat{R} = 0.7$.

For $\hat{\boldsymbol{R}} = \mathbf{0.7}$, $M = 6$, heat transfer coefficient shows monotonic decreasing behaviour as we increase fin height. For $H^* = 0.2$, the value is Nusselt number is highest.

While for the increased number of fins, $M = 18$, Nusselt number shows alternating behaviour for different values of fin height. It firstly decreases and then it starts increasing. For $H^* = 0.6$, it gives its largest value.

From the above two comparisons, we can deduce that for the less number of fins, if we increase the fin height, it reduces the increase in heat transfer. While for more number of fins, longer fins give more optimal results.

5. Conclusion

The results presented in previous sections can be concluded as follows:

- A comparison of present results with the literature results gives the validity proof of the numerical study.

- The number of fin and fin heights are most effective geometrical parameters.

- The influences of geometrical parameters are dependent on each other.

- The location of the regions of high velocities is dependent on geometrical parameters.

- Large velocity gradients exist near the fin tip and outer pipe inner surfaces.

- The value of $\hat{R}$ affects the heat transfer rate significantly, for larger fins and for higher number of fins.

- The value of Nusselt number at $M = 18$, $\hat{R} = 0.5$, gives the optimal value for $H^* = 0.8$.

- For $\hat{R} = 0.7$, the value of Nu(Cu) goes on increasing with higher values of M, at $H^* = 0.8$.

Author details

Ghazala Ashraf[1]*, Khalid S. Syed[2] and Muhammad Ishaq[1]

1 Department of Mathematics, COMSATS University Islamabad, Vehari Campus, Pakistan

2 Centre for Advanced Studies in Pure and Applied Mathematics, Bahauddin Zakariya University, Multan, Pakistan

*Address all correspondence to: ghazalaashraf@ciitvehari.edu.pk

IntechOpen

© 2019 The Author(s). Licensee IntechOpen. This chapter is distributed under the terms of the Creative Commons Attribution License (http://creativecommons.org/licenses/by/3.0), which permits unrestricted use, distribution, and reproduction in any medium, provided the original work is properly cited. CC BY

References

[1] Incropera FP, DeWitt DP. Fundamentals of Heat and Mass Transfer. 4th ed. New York: John Wiley and Sons; 1996

[2] Özisik MN. Heat Transfer-a Basic Approach. McGraw-Hill Book Company; 1985

[3] Nasiruddin MH, Siddiqui K. Heat transfer augmentation in a heat exchanger tube using a baffle. International Journal of Heat and Fluid Flow. 2007;**28**:318-328

[4] Zeitoun O, Hegazy AS. Heat transfer for laminar flow in internally finned pipes with different fin heights and uniform wall temperature. Heat and Mass Transfer. 2004;**40**:253-259

[5] Suryanarayana NV, Apparao TVVR. Heat transfer augmentation and pumping power in double-pipe heat exchangers. 1994;**9**(4):436-444

[6] Li Q, Ma L, Chen Z, Warnecke H-J. Heat transfer characteristics of tube with elliptic pin fins in cross flow of air. Heat and Mass Transfer. 2003;**39**: 529-533

[7] Adegun IK, EL-Suleiman A, Hussein AK. Numerical simulation of forced convective flow and heat transferrin gravitational inclined circular pipes with equi-spaced internal fins. Journal of Basic and Applied Scientific Research. 2013;**3**(10):182-193

[8] Pagliarini G. Effects of axial conduction in the wall and the fluid on the conjugate heat transfer in thick-walled circular tubes. International Communications in Heat and Mass Transfer. 1988;**15**:581-591

[9] Dorfman A, Renner Z. Conjugate problems in convective heat transfer. Mathematical Problems in Engineering. 2009;**27**

[10] Kumar IO. Conjugate problems of heat transfer in laminar boundary layer with injection. Journal of Engineering Physics. 1968;**14**(5):411-416

[11] Barozzi GS, Pagliarani G. A method to solve conjugate heat transfer problems: The case of fully developed laminar flow in a pipe. Journal of Heat Transfer. 1985;**107**:77-83

[12] Mori S, Sakakibara M, Tanimoto A. Steady heat transfer to laminar flow with conduction in the wall tube. Heat Transfer Japanese Research. 1976;**5**(4): 17-25

[13] Sakakibara M, Mori S, Tanimoto A. Conjugate heat transfer with laminar flow in an annulus. The Canadian Journal of Chemical Engineering. 1987; **65**:541-549

[14] Kettner IJ, Degani D, Gutfinger C. Numerical study of laminar heat transfer in internally finned tubes. Numerical Heat Transfer, Part A. 1991; **20**:159-180

[15] Fiebig M, Grosse-Gorgemann A, Chen Y, Mitra NK. Conjugate heat transfer of a finned tube. Part A: Heat transfer behaviour and occurrence of heat transfer reversal. Numerical Heat Transfer, Part A. 1995;**28**:133-146

[16] Nguyen TM, Khodadadi JM, Vlachos NS. Laminar flow and conjugate heat transfer in rib roughened tubes. Numerical Heat Transfer, Part A. 1989; **15**:165-179

[17] Nordstorm J, Berg J. Conjugate heat transfer using modified interface conditions for Navier-Stokes equations. LiTH-MAT-R–2011/18–SE

[18] Sohail M, Fakhir Hasani HS. Conjugate conduction—Convection of finned tube annulus in longitudinal

laminar flow. In: Proceedings of the 4th BSME-ASME International Conference on Thermal Engineering; 27–29 December, 2008

[19] Syed KS, Iqbal M, Mir NA. Convective heat transfer in the thermal entrance region of finned double-pipe. Heat and Mass Transfer. 2007;**43**: 449-457

[20] Iqbal M. Numerical study of laminar heat transfer through a finned double-pipe heat exchanger [PhD thesis]. Multan, Pakistan: Centre for Advanced Studies in Pure and Applied Mathematics (CASPAM), Bahauddin Zakariya University; 2006

Chapter 4

Control Analysis of Building-Integrated Photovoltaic System

Marwa Ben Saïd-Romdhane, Sondes Skander-Mustapha and Ilhem Slama-Belkhodja

Abstract

In this chapter, a photovoltaic system integrated into the building is investigated. The studied structure includes also a battery energy storage system. The overall system is connected to a four-wire AC bus, with the possibility to supply single-phase and three-phase loads. Each equipment is interfaced with a dedicated power converter. This chapter examines the technical operation of all structure components and gives a detailed mathematical study of the DC/AC power converter control in case of two modes, namely, grid connected mode and standalone mode. The investigated control is based on resonant controller. The resonant controller parameters tuning, which is based on the generalized stability margin criterion, is detailed in this chapter. To prove the performance of the proposed control algorithm, several simulation tests developed under PSIM software were performed and then validated by experimental results.

Keywords: photovoltaic systems, battery energy storage system, building microgrid, DC/AC power converter control, standalone mode, grid connected mode

1. Introduction

Nowadays, rooftop photovoltaic systems (PV) and building-integrated photovoltaic (BIPV) systems are becoming well known and commonly used. The growth of these installations is due to their environmental advantages in addition to their social and economic benefits. Indeed, since building electricity consumption accounts for a large proportion of a country's overall consumption, and tends to increase further for the coming years, local generation offers an ideal solution [1–4].

Regarding the obstruction of PV systems' fluctuating aspect, it can be derived in various ways. The integration of battery energy storage system (BESS) is considered as efficient and complementation solution, mainly for standalone microgrids [5]. Urban photovoltaic systems are usually connected to the distribution network, but the operation in standalone is also possible [6, 7].

In order to ameliorate the PV system efficiency, an adequate control strategy should be introduced. In the literature, several control techniques are developed: integral proportional regulators, resonant correctors, hysteresis correctors, sliding mode controls, predictive controls, and so on [8–11].

IntechOpen

This chapter investigates the operation of PV system devoted to building application. It gives an overview of the control of all integrated power converters and then explains in details the control of the DC/AC power converter in both operation modes, namely, standalone mode and grid connected mode. For the grid connected mode, the control must ensure that the AC bus voltage remains within the acceptable range, and for standalone mode the DC/AC converter is controlled to inject generated PV power into the AC-link.

This chapter first outlines overall system description, followed by a review of each power converter control. A detailed mathematical study is dedicated to the DC/AC converter control in grid connected and autonomous modes. Simulation results and experimental validation are subsequently presented.

2. System description

The building solar system structure is given in **Figure 1**. It is composed of a PV panels in parallel with a battery energy storage system which are linked to a DC bus, a DC/AC power converter, and an LCL filter interfacing between DC and AC bus. Single- and three-phase linear and nonlinear loads are connected to the AC bus.

The linear building loads are modeled by a resistive load, and the nonlinear ones are modeled by a rectifier connected to a capacitive filter at the DC side. This model is conformed to many building loads, similar to televisions, personal computers, and fluorescent lamp ballast [12]. In case of three-phase balanced loads, the neutral current is zero, but since several building loads are single phase and include electronic converters, their waves include harmonics which induce a nonzero neutral current. Regarding neutral wire, the more common considered structures are presented in **Figure 2**. The first structure is based on DC-link neutral point where the neutral wire is generated via two identical capacitors (**Figure 2a**). In the second structure, the neutral wire is generated through a Delta/Star grounded transformer as shown in **Figure 2b** [13]. As to the third configuration, it is based on four-leg

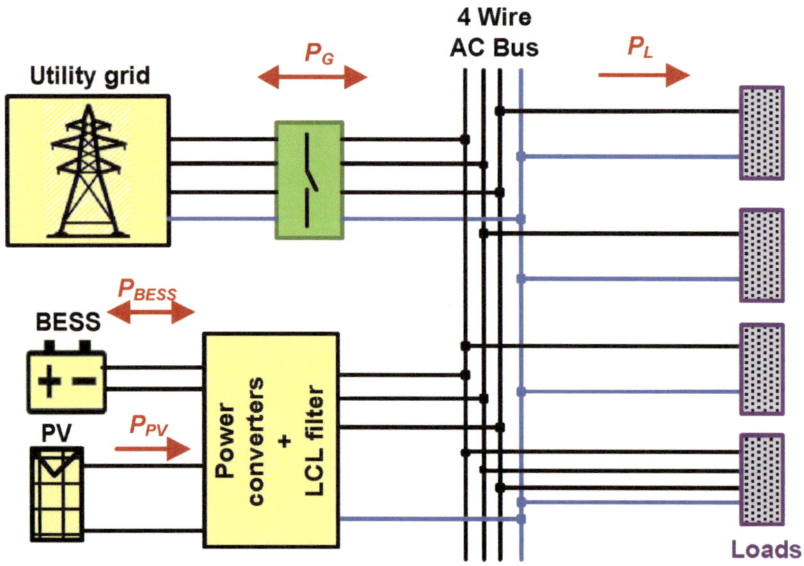

Figure 1.
Photovoltaic system including BESS.

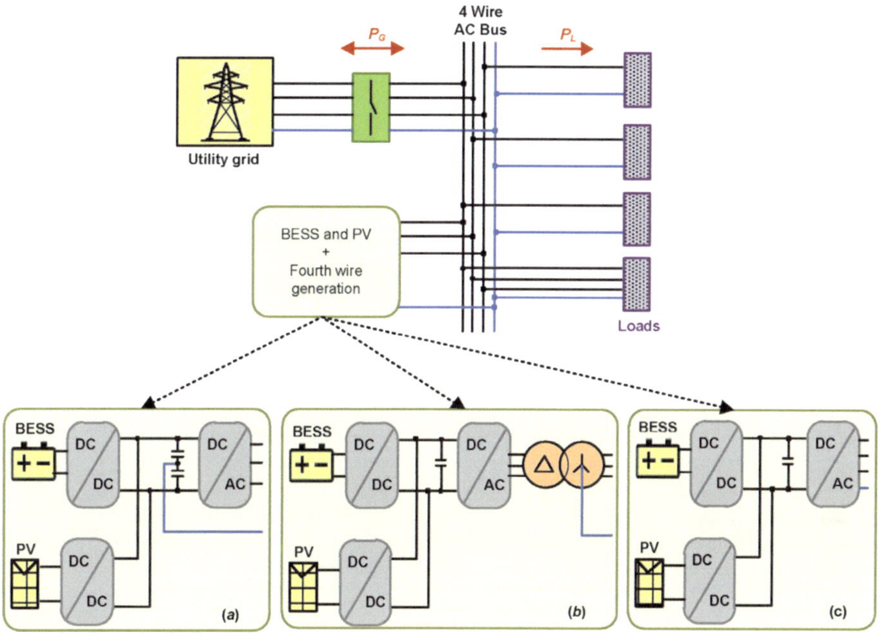

Figure 2.
Different structures that integrate the fourth wire.

power converter (**Figure 2c**) [14–16]. In this chapter the structure with transformer is adopted.

The PV system presents two operating modes according to the grid state:

- Grid connected mode: this mode is activated when the grid is available. In this case, the power surplus is injected into the grid, and if the consumption is superior to local generation, the power flow will be directed from the grid to loads and eventually to charge batteries according to their stat of charge (SOC).

- Standalone mode: this mode is activated when the grid is absent. In this case, building loads are supplied first by the PV system then if necessary by the BESS. In case of power deficit, the shedding of non-priority loads is carried out.

3. Power converter control

An overview of the control of each converter presented in **Figure 2b** is subsequently presented.

3.1 BESS DC/DC converter control

Batteries are frequently integrated to PV systems thanks to their special energy characteristics. Indeed, batteries have a high energy density, which ensure long time of stable operation. The charging time and number of cycles depend on the adopted technology.

The battery power flow is bidirectional. In discharge mode, the power is supplied by battery, and in charging mode, the power is absorbed by battery. For both modes, the state of charge limits should be respected to not affect the battery lifetime.

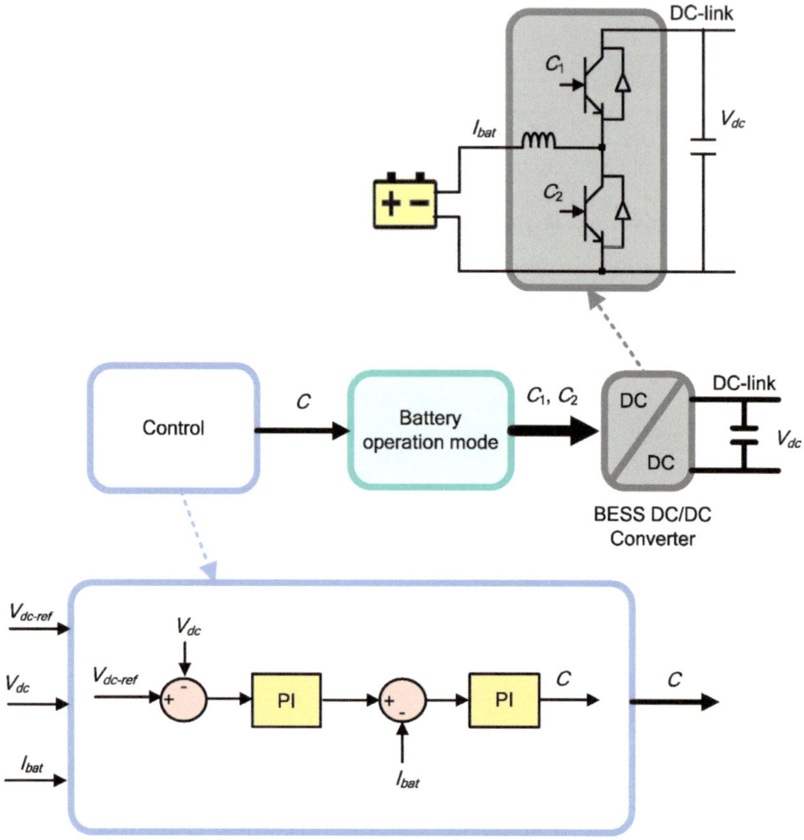

Figure 3.
Control strategy of BESS.

The BESS incorporates a DC/DC power converter that manages battery operation modes according to the appropriate control. A cascade control is adopted, the inner loop regulates the battery current, and the external one regulates the DC-link voltage.

The switches C_1 and C_2 (**Figure 3**) are controlled individually. In case of battery charging, C_1 is controlled and in case of battery discharging, C_2 is controlled.

3.2 PV DC/DC converter control

The structure of the two-stage power conversion is adopted for the PV system; this configuration is commonly privileged in the majority of the PV systems. The difference with the conventional structure is that the V_{dc} regulation is ensured by the BESS. As to the control of the DC/DC converter, it aims to ensure Maximum Power Point Tracking (MPPT) (which corresponds to the peak point of the power versus the voltage curve). In the case of this study, the Perturbation and Observation (P&O) algorithm is applied. The inputs of the P&O algorithm are the solar radiation G and the temperature T as shown in **Figure 4**.

3.3 DC/AC converter control

3.3.1 Modeling of the DC/AC converter

The output of the DC/AC and the LCL filter are modeled in single phase as shown in **Figure 5**. According to this figure, the obtained results are expressed as follows:

DOI: http://dx.doi.org/10.5772/intechopen.91739

$$V_c = V_i - L_1 I_{L1} s \tag{1}$$

$$I_c = I_{L1} - I_{L2} \tag{2}$$

$$V_c = \frac{1}{C_f s} I_c \tag{3}$$

Based on Eq. (3), the transfer function between the current I_{L1} and the voltage $(V_i - V_c)$ is given by the following equation:

$$\frac{I_{L1}}{V_i - V_c} = \frac{1}{L_1 s} \tag{4}$$

According to Eqs. (2) and (3), the transfer function between the voltage V_c and the current $(I_{L1} - I_{L2})$ is expressed as follows:

$$\frac{V_c}{I_{L1} - I_{L2}} = \frac{1}{C_f s} \tag{5}$$

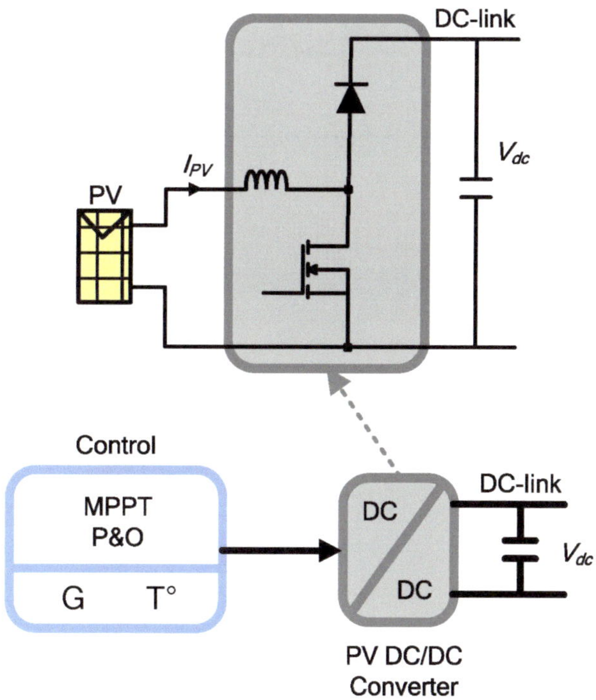

Figure 4.
Control of the PV DC/DC converter.

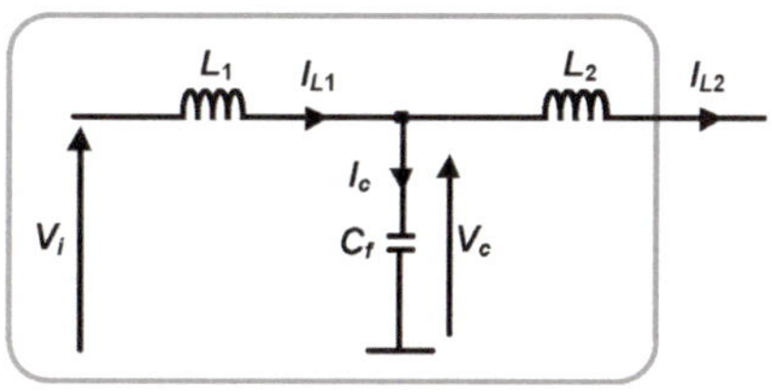

Figure 5.
LCL filter single-phase modeling.

The transfer functions given by Eqs. (4) and (5) allow the deduction of the system block diagram given by **Figure 6**.

3.3.2 Control of the DC/AC converter

3.3.2.1 Standalone mode

In standalone mode, the DC/AC converter control ensures that the LCL filter capacitor voltages are equal to their references. In that case, the converter control includes two cascade loops as shown in **Figure 7**. The external loop is based on a resonant controller RC_1 used to regulate the voltage across the LCL filter capacitor. This loop generates at its output the reference current $I_{c\text{-}ref}$. This current will be added to the current I_2 to provide the inner loop reference current $I_{1\text{-}ref}$. The inner loop is based on a resonant controller, and in this work, it simplified to a simple gain G. In the following, the tuning of the parameters of the voltage external loop and the current inner loop will be presented and detailed.

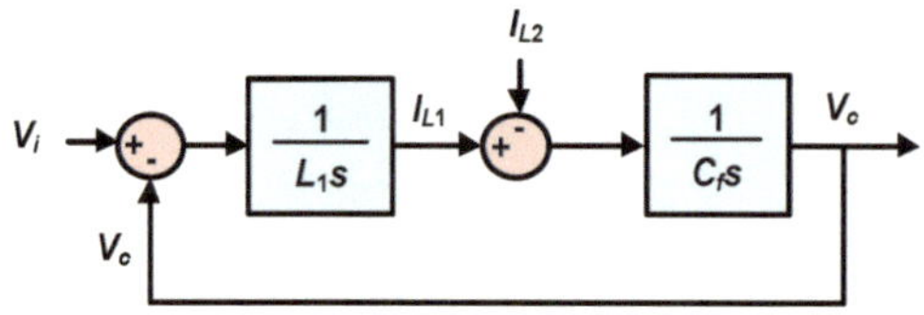

Figure 6.
System filter block diagram.

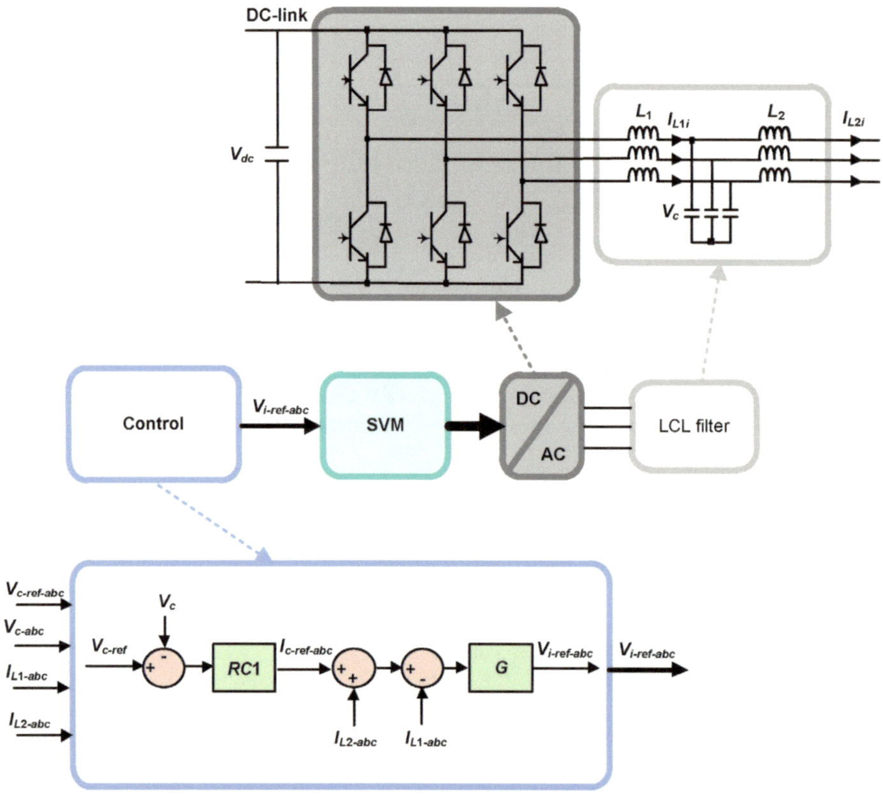

Figure 7.
Control strategy of DC/AC converter in the case of standalone mode.

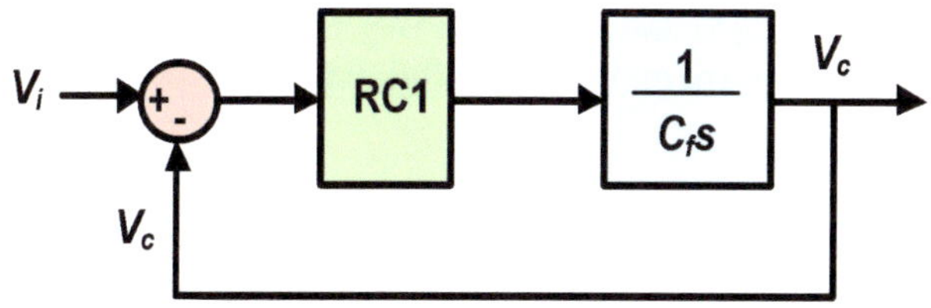

Figure 8.
Block diagram of the external voltage loop.

3.3.2.1.1 Tuning of the external loop resonant controller RC_1

For simplification reasons, it is assumed that the internal current loop is faster than the external voltage loop. Thus, it can be approximated equal to the unity by associating it with the PWM function. The following block diagram is then obtained for the determination of the external voltage loop resonant controller parameters as presented in **Figure 8**.

According to **Figure 4**, the open and closed-loop system transfer functions are expressed by Eqs. (7) and (8), respectively. Note that the transfer function of the resonant controller RC_1 is given by Eq. (6):

$$F_{CR1}(s) = \frac{c_{2c}s^2 + c_{1c}s + c_{0c}}{s^2 + \omega_0^2} \tag{6}$$

$$F_{OL\text{-}Vc}(s) = \frac{V_c}{V_{c\text{-}ref} - V_c} = \frac{c_{2c}s^2 + c_{1c}s + c_{0c}}{C_f s^3 + C_f \omega_0^2 s} \tag{7}$$

$$F_{CL\text{-}Vc}(s) = \frac{V_c}{V_{c\text{-}ref}} = \frac{c_{2c}s^2 + c_{1c}s + c_{0c}}{C_f s^3 + c_{2c}s^2 + \left(C_f \omega_0^2 + c_{1c}\right)s + c_{0c}} \tag{8}$$

The method chosen for the resonant controller parameters tuning is based on the generalized stability margin criterion [17, 18]. The reference polynomial P_{GSMc} defined by this criterion is expressed as follows:

$$P_{GSMc}(s) = \lambda_c(s + r_c)(s + r_c + j\omega_{ic})(s + r_c - j\omega_{ic}) \tag{9}$$

where λ_c, r_c, and ω_{ic} are the factorization coefficient, the abscissa, and the ordinate in the complex plane. On the other hand, the system characteristic polynomial is deduced from Eq. (8), and it is expressed as follows:

$$P_c(s) = C_f s^3 + c_{2c}s^2 + \left(C_f \omega_0^2 + c_{1c}\right)s + c_{0c} \tag{10}$$

According to the generalized stability margin criterion, the resonant controller parameters are tuned by identifying the characteristic polynomial of the closed-loop system $P_c(s)$ with the reference polynomial $P_{GSM}(s)$ as shown in Eq. (11):

$$P_{GSMc}(s) = P_c(s) \tag{11}$$

The identification of $P_{GSM}(s)$ and $P_c(s)$ allows the deduction of the current inner loop resonant controller parameters as shown in the following equation:

$$\begin{cases} c_{2c} = 3r_c\lambda_c \\ c_{1c} = \lambda_c\left(3r_c^2 + \omega_{ic}^2\right) - C_f\omega_0^2 \\ c_{0c} = \lambda_c\left(r_c^3 + r_c\omega_{ic}^2\right) \\ \lambda_c = C_f \end{cases} \tag{12}$$

We choose r_c equal to 200 and ω_{ic} equal to ω_g. For C_f equal to 30 µF, the resonant controller RC_1 parameters are given by the following equation:

$$\begin{cases} c_{2c} = 0.018 \\ c_{1c} = 3.6 \\ c_{0c} = 832.17 \end{cases} \tag{13}$$

For the obtained resonant controller parameters, **Figure 9** shows the pole maps of $F_{CL\text{-}Vc}(s)$. As shown in this figure, the system is stable and the expected stability margin r_c is obtained. **Figure 10** shows the Bode diagram of $F_{OL\text{-}Vc}(s)$. This figure shows that the obtained gain margins G_m and P_m are equal to infinity and 72.8°, respectively. **Figure 11** presents the gain of $F_{CL\text{-}Vc}(s)$ and shows that the bandwidth of the external voltage loop is equal to 24 Hz. It should be noted here that the larger is the bandwidth, the faster is the system.

3.3.2.1.2 Tuning of the inner loop gain G

According to **Figure 6**, the block diagram of the current inner loop is given by **Figure 12**.

According to **Figure 12**, the open and closed-loop transfer functions are given by Eqs. (14) and (15), respectively:

$$F_{OL\text{-}IL1}(s) = \frac{I_{L1}(s)}{I_{L1\text{-}ref}(s) - I_{L1}(s)} = \frac{G}{L_1 s} \tag{14}$$

$$F_{CL\text{-}IL1}(s) = \frac{I_{L1}}{I_{L1\text{-}ref}} = \frac{1}{\frac{L}{G}s + 1} = \frac{1}{1 + \tau_i s} \quad \text{where} \quad \tau_i = \frac{L_1}{G} \tag{15}$$

The inner current loop must ensure a response time much smaller than the external voltage loop. To this purpose, the gain G is selected so that the real part of the inverse of the closed-loop time constant τ_i is greater than the stability margin chosen for the tuning of the voltage external loop (r_c = 200) as shown in Eq. (16).

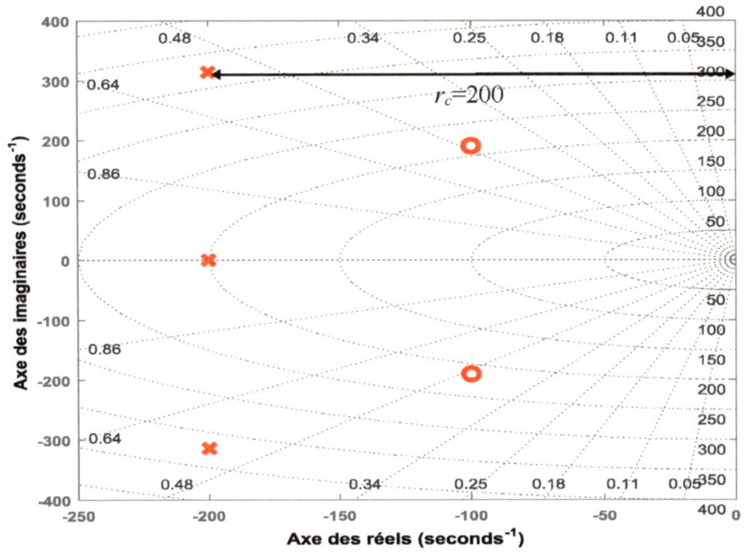

Figure 9.
Pole map of $F_{CL\text{-}Vc}(s)$.

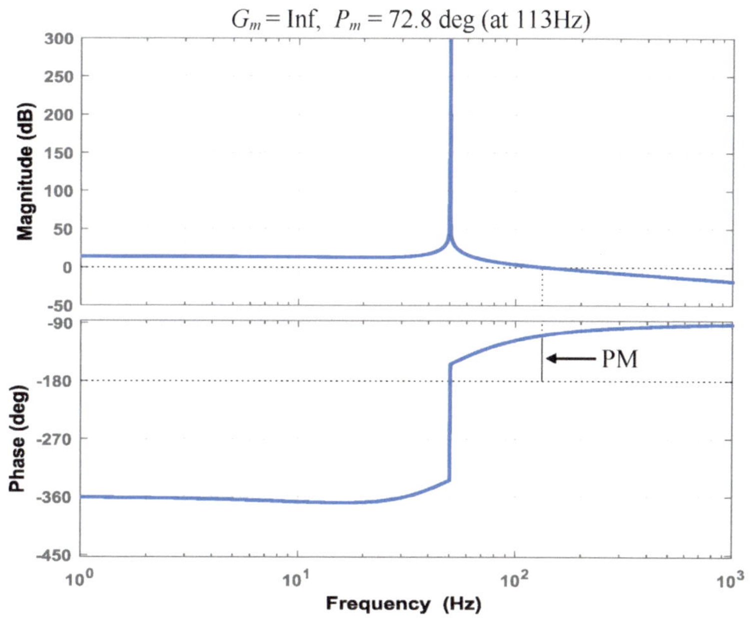

Figure 10.
Bode diagram of $F_{OL\text{-}Vc}(s)$.

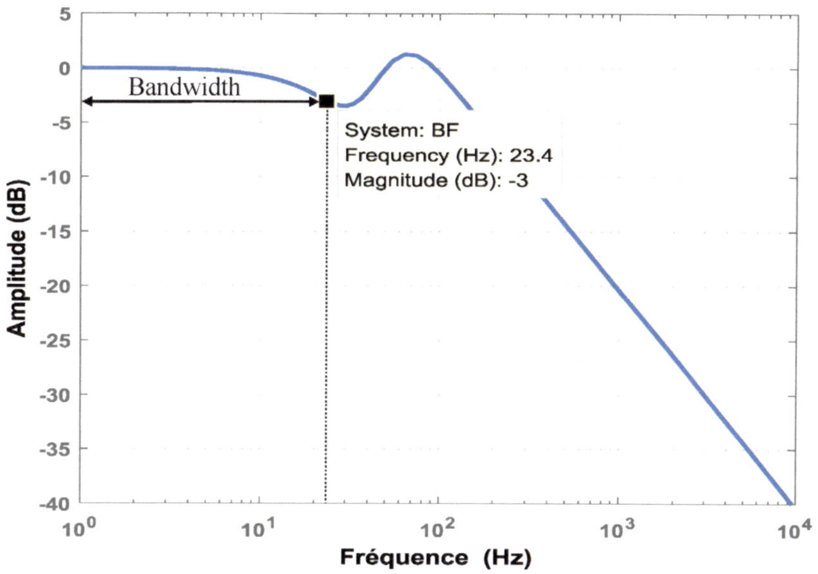

Figure 11.
Bode diagram of $F_{CL\text{-}Vc}(s)$.

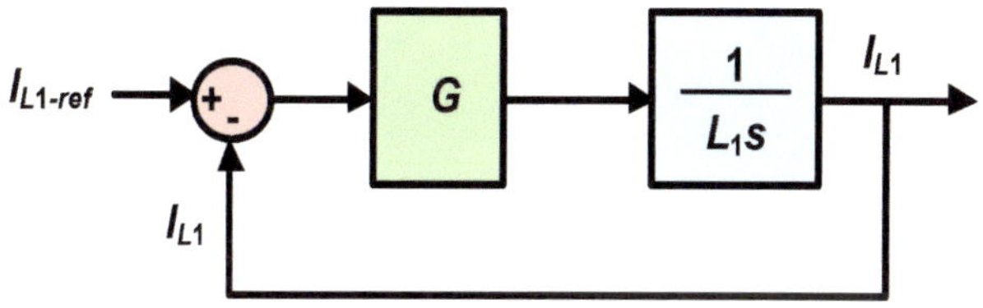

Figure 12.
Block diagram of the inner current loop.

$$\frac{1}{\tau_i} \gg 100 \quad \Rightarrow \quad \frac{G}{L_1} \gg 100 \quad \Rightarrow \quad G \gg 0.1 \tag{16}$$

We select G equal to 10. For this value, **Figures 13** and **14** present the pole maps of $F_{CL\text{-}IL}(s)$ and the Bode diagram of $F_{OL\text{-}IL}(s)$, respectively. These figures show that

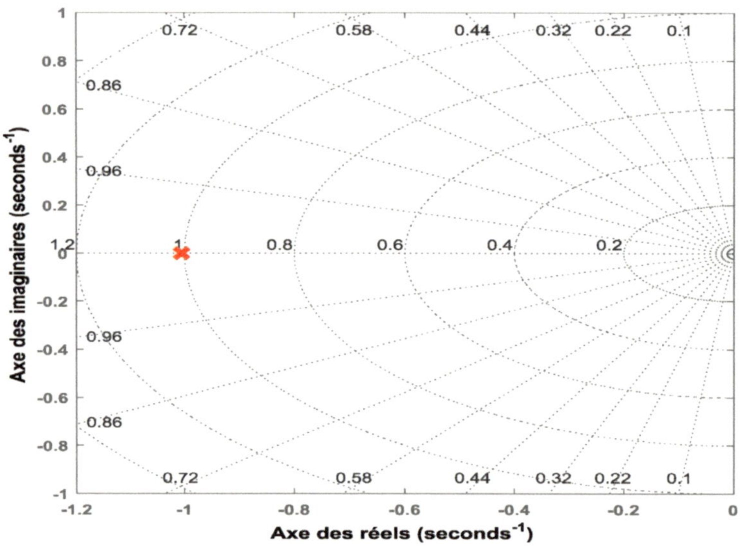

Figure 13.
Pole map of $F_{CL\text{-}IL}(s)$.

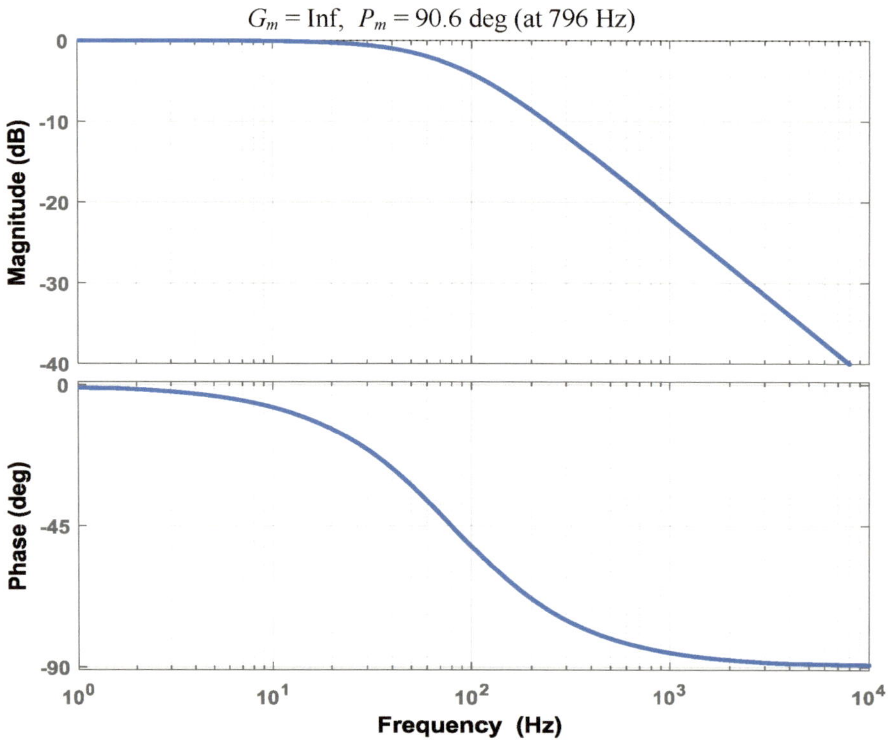

Figure 14.
Bode diagram of $F_{OL\text{-}IL}(s)$.

the system is stable and the obtained G_m is equal to infinity and the P_m is equal to 90.3°. **Figure 15** presents the gain of $F_{CL\text{-}IL}(s)$ and shows that the bandwidth of the inner current loop is equal to 785 Hz. This value is much higher than the bandwidth of the voltage external loop and shows that the current inner loop is much faster than the voltage external loop.

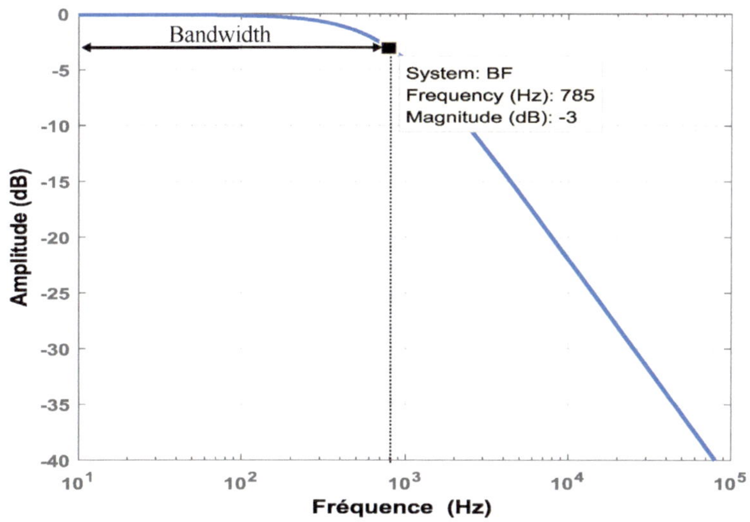

Figure 15.
Bode diagram of $F_{CL\text{-}IL}(s)$.

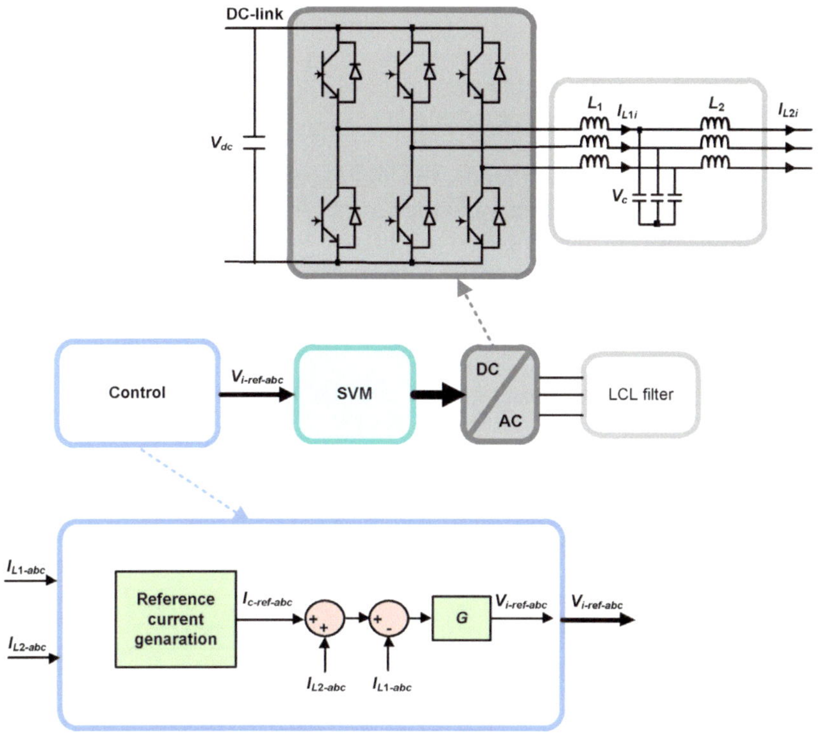

Figure 16.
Control strategy of DC/AC converter in case of grid connected mode.

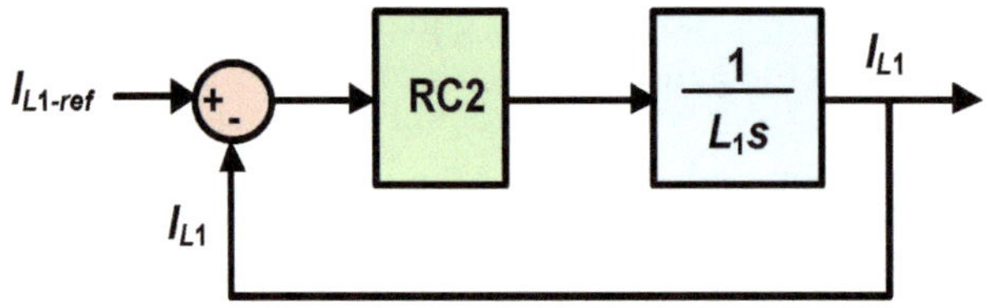

Figure 17.
Block diagram of the current loop.

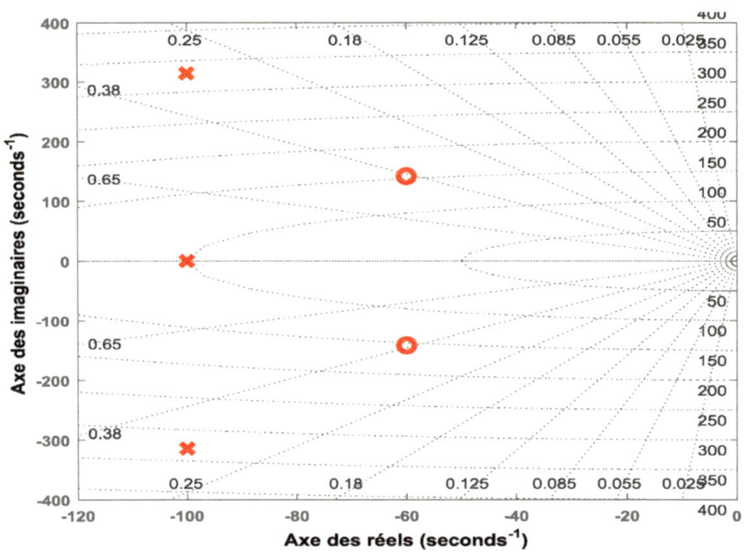

Figure 18.
Pole map of $F_{CL\text{-}i}(s)$.

3.3.2.2 Grid connected mode

In grid connection operation mode, the DC/AC converter controls power exchange with grid. In this mode, only the current loop is controlled. This loop is based on a resonant controller as shown in **Figure 16**.

The simplified block diagram of the current loop is given by **Figure 17**.

Based on **Figure 17**, the open and closed-loop transfer functions are given by Eqs. (18) and (19), respectively. The transfer function of the resonant controller RC_2 is given by Eq. (17):

$$F_{CR2}(s) = \frac{i_{2i}s^2 + i_{1i}s + i_{0i}}{s^2 + \omega_0^2} \tag{17}$$

$$F_{OL\text{-}i}(s) = \frac{I_{L1}}{I_{L1\text{-}ref} - I_{L1}} = \frac{i_{2i}s^2 + i_{1i}s + i_{0i}\omega_0^2}{L_1 s^3 + L_1\omega_0^2 s} \tag{18}$$

$$F_{CL\text{-}i}(s) = \frac{I_{L1}}{I_{L1\text{-}ref}} = \frac{i_{2i}s^2 + i_{1i}s + i_{0i}}{L_1 s^3 + i_{2i}s^2 + \left(L_1\omega_0^2 + i_{1i1}\right)s + i_{0i1}} \tag{19}$$

For the tuning of the internal loop resonant controller, the generalized stability margin criterion is considered. The system characteristic polynomial $P_i(s)$ is deduced from Eq. (19), and it is expressed as follows:

$$P_i(s) = L_1 s^3 + i_{2i} s^2 + \left(L_1 \omega_0^2 + i_{1i}\right)s + i_{0i} \tag{20}$$

The identification between the system characteristic polynomial $P_i(s)$ and the generalized stability margin criterion reference polynomial $P_{GSMi}(s)$ [Eq. (21)] and the resonant controller RC_2 parameters are deduced as in Eq. (22):

$$P_{GSMi}(s) = \lambda_i (s + r_i)(s + r_i + j\omega_{ii})(s + r_i - j\omega_{ii}) \tag{21}$$

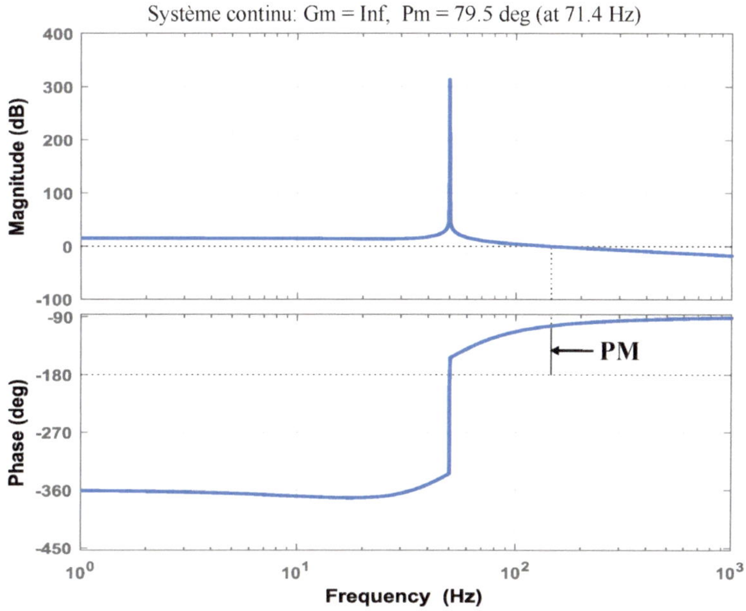

Figure 19.
Bode diagram of $F_{OL\text{-}i}(s)$.

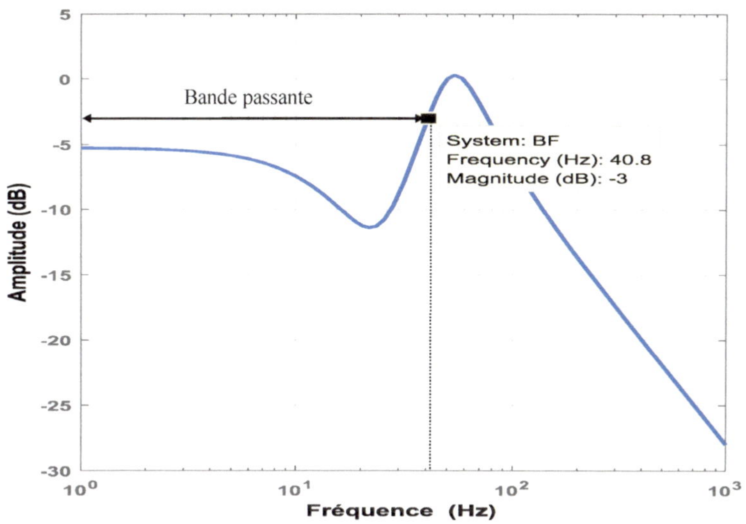

Figure 20.
Bode diagram of $F_{CL\text{-}i}(s)$.

$$\begin{cases} i_{2i} = 3r_i\lambda_i \\ i_{1i} = \lambda_i\left(3r_i^2 + \omega_{ii}^2\right) - L_1\omega_0^2 \\ i_{0i} = \lambda_i\left(r_i^3 + r_i\omega_{ii}^2\right) \\ \lambda_i = L_1 \end{cases} \tag{22}$$

We choose r_i equal to 100 and ω_{ii} equal to ω_g. For L_i equal to 2 mH, the resonant controller RC_2 parameters are given by the following equation:

$$\begin{cases} i_{2i} = 0.5 \\ i_{1i} = 60 \\ i_{0i} = 11870 \end{cases} \tag{23}$$

For the obtained resonant controller parameters, **Figure 18** shows the pole map of $F_{CL\text{-}i}(s)$. Based on this figure, the stability margin r_i is equal to the desired one. **Figure 19** gives the bode diagram of $F_{OL\text{-}i}(s)$. As mentioned on this figure, the gain margins G_m and P_m are equal to infinity and 79.5°, respectively. **Figure 20** presents the gain of $F_{CL\text{-}i}(s)$ and shows that the bandwidth of the internal current loop is equal to 40.8 Hz.

4. Simulation results

Several simulation tests developed under PSIM software were done. **Figure 21** presents the LCL filter capacitor voltage in islanded mode for different values of voltage reference $V_{c\text{-}ref\text{-}abc}$. As shown in this figure, the obtained voltages are equal to their references. **Figure 22** presents the power injected into the AC bus during 24 hours; this power corresponds to the PV generation. In this case all the batteries are considered to be charged to their SOC_{max}. The deduced reference current is presented in **Figure 23**.

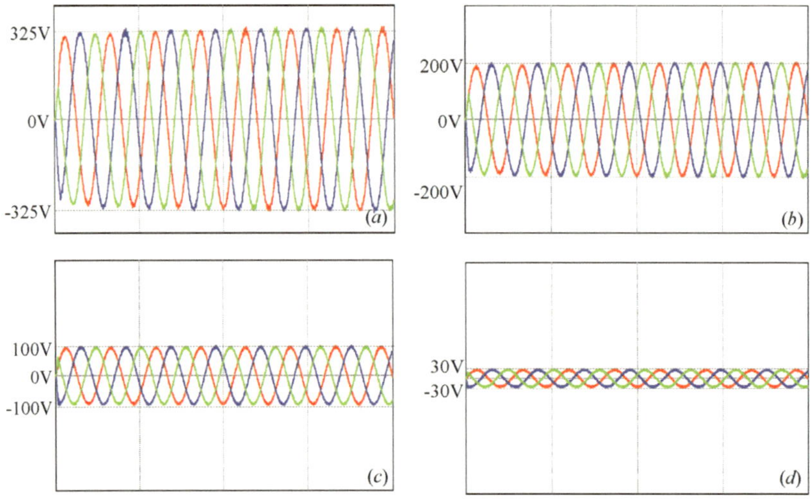

Figure 21.
LCL filter capacitor voltage in islanded mode for different values of voltage reference $V_{c\text{-}ref\text{-}abc}$*:*
(a) $V_{c\text{-}ref\text{-}abc}$ *= 325 V, (b)* $V_{c\text{-}ref\text{-}abc}$ *= 200 V, (c)* $V_{c\text{-}ref\text{-}abc}$ *= 100 V, and (d)* $V_{c\text{-}ref\text{-}abc}$ *= 30 V.*

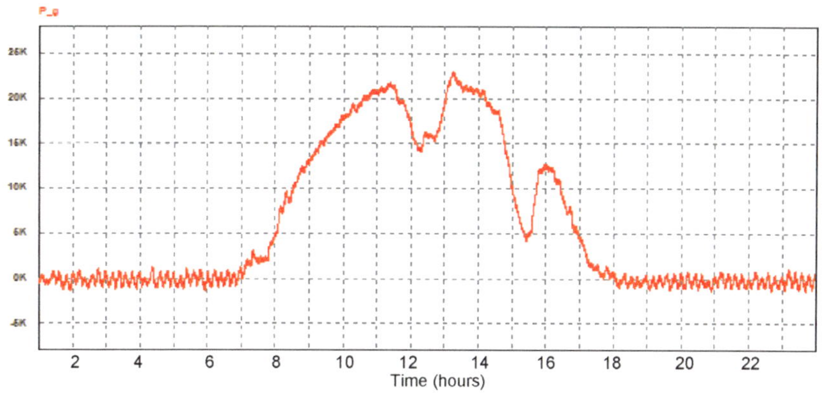

Figure 22.
Power injected into AC bus in case of connected mode.

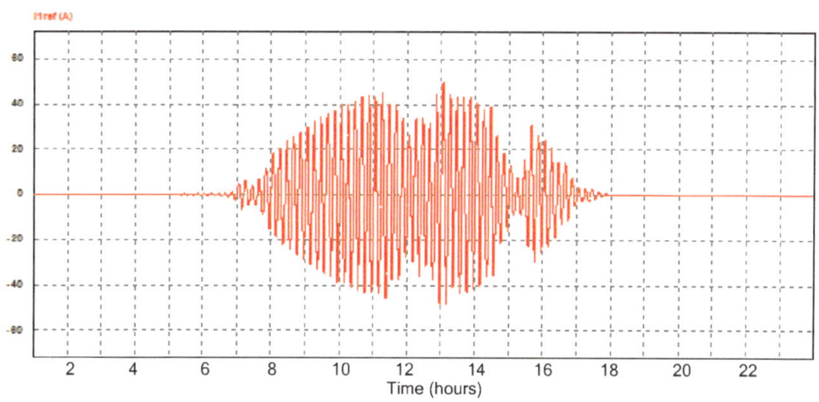

Figure 23.
Reference current in case of connected mode.

5. Experimental results

Figure 24 shows the experimental test bench. The used AC/DC converter is from SEMIKRON. Currents and voltages are censored via LEM LV25 and LEM 55LP, respectively, as given in **Figure 24**. The control algorithm is implemented on the STM32F4 Discovery. The acquisition time is set to 100 μs. **Figure 25** presents the LCL filter capacitor voltage in islanded mode for different values of voltage reference $V_{c\text{-}ref\text{-}abc}$. As shown in this figure, the obtained voltages are equal to their references.

6. Conclusion

In this chapter, the control of power converters integrated in building solar system is investigated. The studied system is composed of a PV panel in parallel with a battery energy storage system which are linked to a DC bus, a DC/AC power converter, and an LCL filter interfacing between DC and AC bus. Single- and three-phase linear and nonlinear loads are connected to a four-wire AC bus.

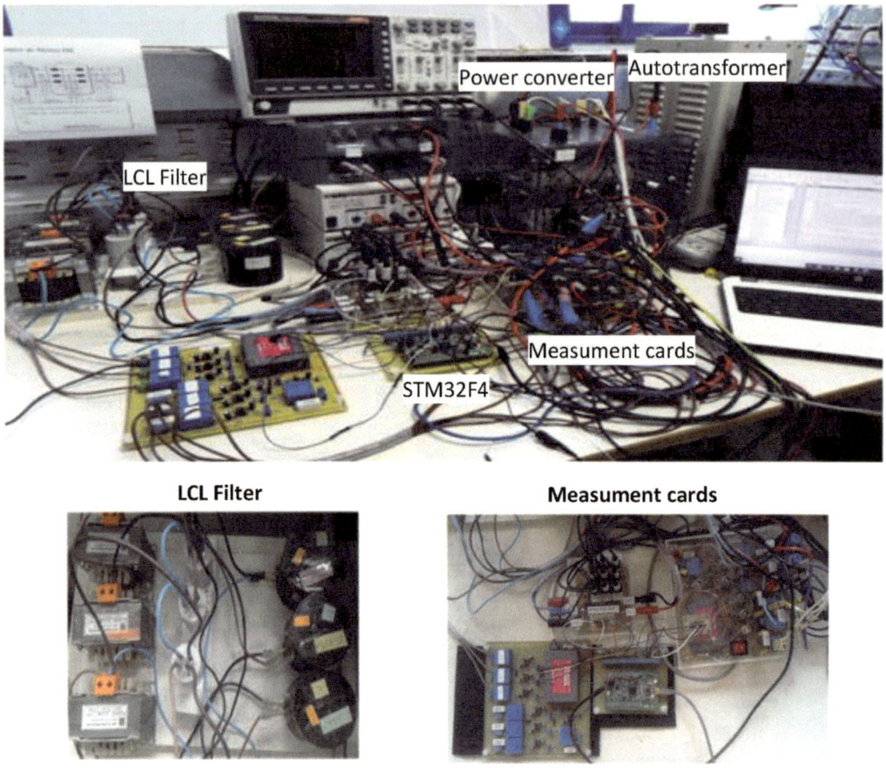

Figure 24.
Experimental test bench.

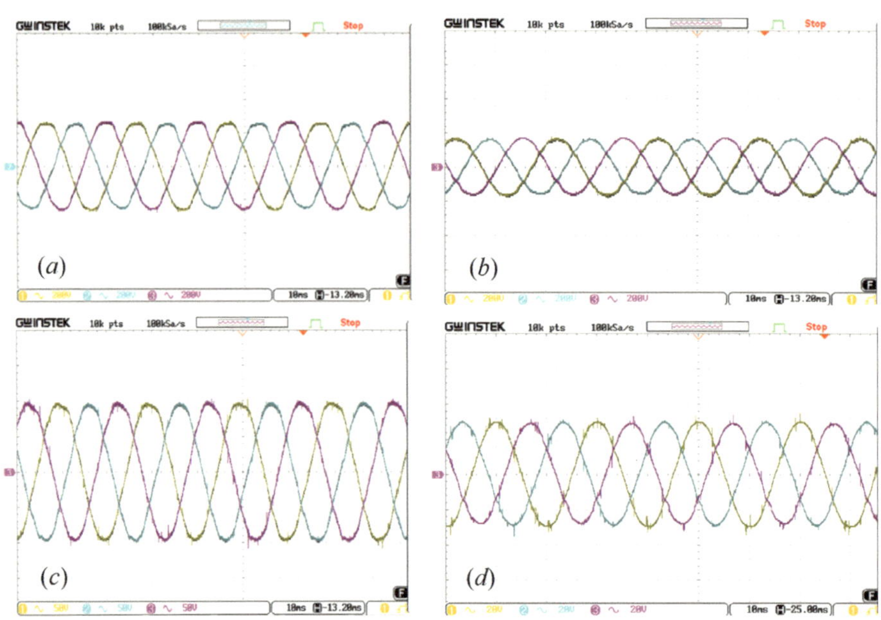

Figure 25.
Experimental results for LCL filter capacitor voltage in islanded mode for different values of voltage reference $V_{c\text{-ref-abc}}$*: (a)* $V_{c\text{-ref-abc}} = 325$ *V, (b)* $V_{c\text{-ref-abc}} = 200$ *V, (c)* $V_{c\text{-ref-abc}} = 100$ *V, and (d)* $V_{c\text{-ref-abc}} = 30$ *V.*

The neutral wire is generated through a Delta/Star grounded transformer. An overview of the control of each power converter is presented. This chapter focuses on the control of the DC/AC power converter. The resonant controller is adopted. A set of simulation and experimental tests were done to show the efficiency of the studied control algorithm.

Acknowledgements

This work was supported by the Tunisian Ministry of Higher Education and Scientific Research under Grant LSE-ENIT-LR11ES15.

Author details

Marwa Ben Saïd-Romdhane[1*], Sondes Skander-Mustapha[1,2] and Ilhem Slama-Belkhodja[1]

1 Université de Tunis El Manar, Ecole Nationale d'Ingénieurs de Tunis, Laboratoire des Systèmes Electriques, Tunis, Tunisie

2 Université de Carthage, Ecole Nationale d'Architecture et d'Urbanisme, Tunisie

*Address all correspondence to: marwa.bensaidromdhane@enit.utm.tn

IntechOpen

© 2020 The Author(s). Licensee IntechOpen. This chapter is distributed under the terms of the Creative Commons Attribution License (http://creativecommons.org/licenses/by/3.0), which permits unrestricted use, distribution, and reproduction in any medium, provided the original work is properly cited.

References

[1] Holland I, Doorsamy W, Nixon K. Analysis of lightning surge effects on small-scale rooftop photovoltaic systems. In: 2018 Power Systems Computation Conference (PSCC), Dublin; 2018. pp. 1-6. DOI: 10.23919/PSCC.2018.8443046

[2] Fina B, Fleischhacker A, Auer H, Lettner G. Economic assessment and business models of rooftop photovoltaic systems in multiapartment buildings: Case studies for Austria and Germany. Journal of Renewable Energy. 2018; **2018**:1-16. DOI: 10.1155/2018/9759680

[3] Saker N, Al-Qattan A, Al-Otaibi A, Al-Mulla A. Cost–benefit analysis of rooftop photovoltaic systems based on climate conditions of gulf cooperation council countries. IET Renewable Power Generation. 2018;**12**(9):1074-1081. DOI: 10.1049/iet-rpg.2017.0309

[4] Dehwah AHA, Asif M. Assessment of net energy contribution to buildings by rooftop photovoltaic Systems in hot-humid climates. Renewable Energy. 2018;**131**:1288-1299. DOI: 10.1016/j.renene.2018.08.031

[5] Ahmadi M, Lotfy ME, Shigenobu R, Howlader AM, Senjyu T. Optimal sizing of multiple renewable energy resources and PV inverter reactive power control encompassing environmental, technical, and economic issues. IEEE Systems Journal. 2019;**13**(3):3026-3037. DOI: 10.1109/JSYST.2019.2918185

[6] O'Shaughnessy E, Cutler D, Ardani K, Margolis R. Solar plus: A review of the end-user economics of solar PV integration with storage and load control in residential buildings. Applied Energy. 2018;**228**:2165-2175. DOI: 10.1016/j.apenergy.2018.07.048

[7] Lu Y, Wang S, Shan K. Design optimization and optimal control of grid-connected and standalone nearly/net zero energy buildings. Applied Energy. 2015;**155**:463-477. DOI: 10.1016/j.apenergy.2015.06.007

[8] Jamroen C, Pannawan A, Sirisukprasert S. Battery energy storage system control for voltage regulation in microgrid with high penetration of PV generation. In: 2018 53rd International Universities Power Engineering Conference (UPEC), Glasgow; 2018. pp. 1-6. DOI: 10.1109/UPEC.2018.8541888

[9] Bag A, Subudhi B, Ray P. An adaptive variable leaky least mean square control scheme for grid integration of a PV system. IEEE Transactions on Sustainable Energy. 2019. DOI: 10.1109/TSTE.2019.2929551

[10] Al-Dhaifallah M, Nassef AM, Rezk H, Nisar KS. Optimal parameter design of fractional order control based INC-MPPT for PV system. Solar Energy. 2018;**159**:650-664. DOI: 10.1016/j.solener.2017.11.040

[11] Hamrouni N, Jraidi M, Ghobber A, et al. Control approach of a connected PV system under grid faults. Electrical Engineering. 2018;**100**:1205. DOI: 10.1007/s00202-017-0560-0

[12] Antunes AF, Baptista JR, Moura AM, Pomilio JA. Study of harmonic distortion in a residential and commercial LV power system. In: 11th International Conference on Electrical Power Quality and Utilisation; 2011. DOI: 10.1109/epqu.2011.6128873

[13] Mohammadi J, Badrkhani Ajaei F, Stevens G. Grounding the AC microgrid. IEEE Transactions on Industry Applications. 2018;**55**:1-1. DOI: 10.1109/tia.2018.2864106

[14] Naderipour A, Abdul-Malek Z, Ramachandaramurthy VK, Kalam A, Miveh MR. Hierarchical control strategy for a three-phase 4-wire microgrid

under unbalanced and nonlinear load conditions. ISA Transactions. 2019;**94**: 352-369. DOI: 10.1016/j.isatra.2019.04.025

[15] Kim G-H, Hwang C, Jeon J-H, Ahn J-B, Kim E-S. A novel three-phase four-leg inverter based load unbalance compensator for stand-alone microgrid. International Journal of Electrical Power & Energy Systems. 2015;**65**:70-75. DOI: 10.1016/j.ijepes.2014.09.035

[16] Sadeghkhani I, Hamedani Golshan ME, Mehrizi-Sani A, Guerrero JM. Low-voltage ride-through of a droop-based three-phase four-wire grid-connected microgrid. IET Generation Transmission and Distribution. 2018;**12**(8):1906-1914. DOI: 10.1049/iet-gtd.2017.1306

[17] Kuperman A. Proportional-resonant current controllers design based on desired transient performance. IEEE Transactions on Power Electronics. 2015;**30**(10):5341-5345. DOI: 10.1109/TPEL.2015.2408053

[18] Pereira LFA, Bazanella AS. Tuning rules for proportional resonant controllers. IEEE Transactions on Control Systems Technology. 2015;**23**(5):2010-2017. DOI: 10.1109/TCST.2015.2389655

Section 2

Computer Simulation

Chapter 5

Computer Simulation and the Practice of Oral Medicine and Radiology

Saman Ishrat, Akhilanand Chaurasia and Mohammad Husain Khan

Abstract

The practice of Oral Medicine and Radiology has long been considered an art form. Collecting and collimating the enormous amount of information each patient brings has always tested the best of our abilities as diagnosticians. However, as the tide of smartphones, cheaper data access, and automation rises, it threatens to wash away all that we have held sacrosanct about conventional clinical practices. In this tussle between what is traditional and what is tantalizing, it is time to question, as diagnosticians, how much can we accede to the invasion of algorithms. How does computer simulation affect the practice of diagnosis in the field of Oral Medicine and Radiology?

Keywords: computer simulation, artificial intelligence, diagnosis

1. Introduction

We have come a long way since the time our ancestors rolled off the first round piece of wood and invented a wheel. Over the years that tiny stroke of luck has rolled on and become a vastly sophisticated industry that demands state-of-the-art technology, and even political and social changes to keep humanity on the move. Today, developments in one field set up ripples that affect the development of other fields. Like seismic waves, the repercussions of that change are felt all over. So, in a time defined largely by software, it is no surprise that technology today is the next second away and about to disrupt the practice of our profession. As with any change, it remains to be seen whether the growing dependence on computers will be for better or for worse.

2. Defining the diagnosis

According to Miller [1], diagnosis is way more than connecting the name of a disease or syndrome with the findings for a patient. It is a recurring process in which the details of a patient such as history, symptoms, signs and how the disease process has unfolded over time, and eventually how that process affects the patient's life, count [1]. The diagnosis of an individual involves a series of information which includes history, symptoms, physical exams, laboratory tests and clinical image interpretations which potentially coincides with the etiology of the patient's illness. The diagnosis of some of the diseases may involve the response of an individual to

IntechOpen

"therapeutic intervention" and the response is studied to show characteristics of a particular disease [2]. For example, the dentist might prescribe antibiotic for 1 week to a patient with decayed tooth and tooth pain. The antibiotic resolves the infection and pain which shows that the patient had secondary infection going on in the carious tooth. The ultimate feature in the diagnosis of disease is to see if the planned therapy is working or if the disease is getting better or getting worse over a period of time, and have there been considerable side effects to the therapy [2]. Evaluation of the diagnoses is not only limited to living individuals. Post-mortem and autopsy examinations have proved to be a great tool in designing the diagnostic parameters. Artificial intelligence and computer learning have cashed upon the complex series of diagnosis in these days to assist humans in coming up with a module for diagnostic processes. In the current scenario, the possibility that artificial intelligence (abbreviated henceforth as AI) and computer aided systems may supersede humans in the diagnostic process, is an impending reality [1].

3. Factors influencing the diagnosis

The diagnostic process is considered to be a complex transition process which begins with the illness of the patient and ends into a result which serves as a data for reference that can be categorized. The diagnoses start from the doctor asking the patient about signs and symptoms of a disease/illness. A doctor's treatment specific to a patient's symptoms and its outcome is important for both the doctor and the patient to see the efficacy of the treatment provided.

The diagnoses are an amalgamation of various processes that broadly depends on three main factors:

3.1 Doctor's knowledge

> *"The eye sees all, but the mind shows us what we want to see."*
>
> *-Shakespeare.*

The doctor's knowledge serves as data set for the human brain to process the likes and similarities from the previous acquired data. This helps in differentiating the two processes. And thus helps to form an opinion regarding a process. This is really critical for diagnosis.

For example, while diagnosing a Central Giant Cell Granuloma of the jaw, the doctor must also be aware of the differential diagnoses, and their appearance to rule those out.

3.2 Doctor's experience

The experience of a doctor plays an important role in the diagnosis of a disease. These experiences contribute in enriching the quality of the data and help in refining the data set and recreating subsets in the data. It is here that heuristics step into the picture. These subsets help in simplifying the data and make it comprehensible. The experiences can also be governed by the amount of different cases seen by the doctor, which in turn is greatly influenced by the geo-fencing of the same.

For example, the diseases that are predominant in some areas of the world, like Lyme's disease in the North Eastern American region, the physicians practicing there will have more experience of those cases and will form a more accurate diagnosis in comparison to the physicians in any other part of the world. Basically the age old dictum at work here is that—if you hear hooves think horses, not zebras.

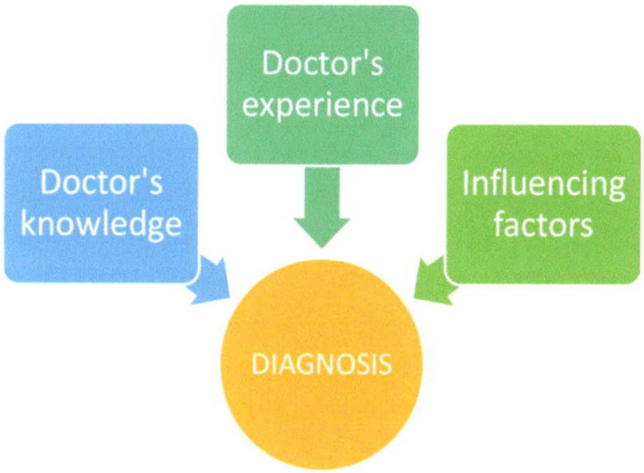

Figure 1.
Diagnostic paradigm.

3.3 Influencing factors

Human brain is a complex structure which plays a pivotal role in making a diagnosis. According to Charles Sherigton (c. 1920), some of the deepest mysteries facing science in the twenty-first century concern the higher functions of the central nervous system: perception, memory, attention, learning, language, emotion, personality, social interaction, decision-making, motor control, and consciousness. Nearly all psychiatric and many neurological disorders are characterized by a dysfunction in the neural systems that mediate these neural processes. In fact, all aspects of human behavior and hence human society are controlled by the human brain: economics and decision making, moral reasoning and law, arts and esthetics, social and global conflict, politics and political decision making, marketing and preference, etc. These functions are greatly altered by the level of stress and the mood that the person has. Thus, a diagnosis is also greatly influenced by the doctor's state of mind, and stress level (**Figure 1**).

4. Computer simulation: understanding AI (artificial intelligence) in computer-aided diagnosis (CAD)

AI and how to use it in CAD has become one of the hottest research topics in medical radiology both in imaging and diagnostics. Although, research in CAD is pretty much established and growing but most radiologists do not as yet, use CAD in their daily routine. The basics of AI and how to use it in CAD for detection and for quantification is defined by the various requirements such as performance, regulatory compliance, reading time reduction and cost efficiency are even today not as sophisticated/dependable as the human mind. Overall the performance of the CAD systems is still a major bottleneck for adaption. However, the usual machine learning and AI strategy can be used to improve CAD by using past and public databases for training and validation. This will create cognitive AI that will help tackle corner cases in CAD and eventually create superior algorithms [3].

Yet all said and done, there is a global consensus that the advent of computer simulation is a crisis in the making for radiology. Not only has the number of imaging studies gone up, but also the number of images per study has drastically increased [4]. Radiology is becoming a victim of its own success, i.e., the disparity and the gap between the overall workload and the number of radiologists has increased dramatically which has resulted in a cost increase. Therefore, new solutions are needed to handle the spurt of data and workload. Computer simulation may be the answer to this evident problem. The key is to use AI and CAD to quicken the diagnostic process and minimize diagnostic errors.

Although, CAD is far much more than just a detection tool but CAD is now widely used as a general term for detection that includes aided extraction of quantitative data from radiology images. A very interesting fact about the algorithm development is that detection and quantification both use the same underlying principle for algorithm creation.

The overall growth in computer simulation or CAD is driven by Moore's law, i.e., the computational power doubles every 2 years [5]. This has been true for the last 5 decades and should continue for at least another decade. Futurists like Kurzweil [6] talk about singularity of AI, i.e., within a decade, a $1000 computer will have the computational strength of a human brain and eventually the power of hundreds of human brains by 2040. The availability of cheaper and faster hardware has allowed for quicker computations and bigger and cleaner databases for algorithm training. All this has led to quicker and better CAD performance and results.

5. History of computer simulation and artificial intelligence (AI)

In the 1980s, the Kurt Rossmann Laboratories for Radiologic Image Research in the Department of Radiology at the University of Chicago first started the systematic research in developing and designing the CAD systems for the diagnosis of the diseases. (Computer-Aided Diagnosis in Medical Imaging: Historical Review, Current Status and Future Potential). Before this there was a significant amount of studies and researches going on in the picture archiving and communication system (PACS) [7–25].

As a matter of fact, the PACS were useful in storing the pictures and reducing the cost for the storage to the hospitals but at that time, it was not thought that the stored pictures, nowadays referred to the data, might be of any clinical significance to the doctors or the clinicians? The storage was one of the fringe benefits of PACS, but the major value addition was the formation of a sample set or data. The researchers started thinking how this data could help the doctors in diagnostic process. This led to the theory of computer-aided diagnosis (CAD) and artificial intelligence (AI) led diagnosis [26].

6. Rise of artificial intelligence (AI) in healthcare: transformative future

The sophistication of artificial intelligence (AI) in doing what humans do has increased by leaps and bounds in the last one decade. In 2019, AI is a fact today and we have seen a shift in the conversation. We are no longer answering the question—what is AI? Today, the primary concern is answering the question—how can we utilize the plethora of information to replicate the human actions in a more efficient and faster way? No other sector is answering this question better than Healthcare. Artificial intelligence and computer simulations are no longer a novelty and if things progress the way they are, these may soon be the norm.

The use cases for AI in healthcare are vast and ever evolving. Just like AI has become a seminal part of our daily lives, AI is also transforming our healthcare ecosystem. When AI is applied strategically to this ecosystem, it not only has the ability to deeply impact the way healthcare is delivered but also how that the healthcare impacts the overall cost structure.

7. Innovation curve

Today, AI has pushed innovation in healthcare to the next level by combining the training data sets with cognitive computing to draw new insights and correlations. This has been possible because of predictive capabilities, complex algorithms and analytics to deliver real time data that is clinically relevant, i.e., transforming healthcare in new ways.

First and foremost, is to help people stay healthy and eventually reducing the frequency of patient and doctor interaction. The new health apps encourage people to live a healthy lifestyle. AI equips healthcare professionals to better understand everyday health patterns and the needs of their patients. The increased use of consumer hardware technologies such as Apple Watch and other medical devices combined with AI is used in pilot projects such as detecting early-stage heart disease. Thus, helping healthcare professionals to better detect and monitor underlying life-threatening events at early, more treatable stages. With more and more money being invested in projects like Apple Health and Common Health of Android platform, the upheavals in how we practice as diagnosticians, are going to be tectonic.

Recently, life threatening diseases such as cancer are being detected more accurately by AI in their early stages. Based on the study done by American Cancer Society, a large proportion of mammograms eventually result in false positives, i.e., 1 in 2 healthy women are diagnosed with cancer when they have none. Using AI in review and translation process of mammograms may help to avoid unnecessary biopsies.

8. Simulation and requirements for artificial intelligence (AI)-aided systems

AI has to meet several demands to be used widely in clinical practice. The major four requirements that we think, are of paramount importance for AI guiding computer simulations, to be helpful in the field of Oral Radiology, are:

a. *AI should improve radiologists' performance*—which means that efficacy and the accuracy of detecting the aberrancy in the scan should be picked up by the AI system [3].

b. *AI should save time*—it should save the radiologist in detecting and diagnosing a disease. One of the important factors which adds efficacy of a system for any machine is that it reduces time. If an AI system is not decreasing the time for the diagnosis process then it is not helping the radiologist at its 100% [3].

c. *AI must be seamlessly integrated into the workflow*—The AI system should be a part of the diagnosis process without being a process in itself. It should make the diagnosis process easy and viable to the doctors [3].

d. *AI should not impose liability concerns*—The AI system used should be HIPAA compliant system and there should have a foolproof close system to prevent any data breach [3].

Most AI systems, and therefore computer simulations used in diagnostics that are based on these, today do not meet all requirements, and this is why most applications described in the rapidly growing body of scientific literature on CAD are not widely used in clinical practice.

9. Artificial intelligence (AI) for lesion detection

Presently, the existing CAD systems are pushed as complementary tools for radiologists to further evaluate certain images that need attention. CADs have a limitation though, i.e., it does not detect all potential lesions and would limit the radiologist to focus only on the areas that the CAD system has identified. Therefore, it is imperative that the radiologist does the evaluation of the complete image. But the CAD system can help detect lesions that the radiologist might have missed [27–41].

10. The heart of computer simulation: working of artificial intelligence (AI) system

It is very complicated and difficult for computers to decipher radiologic images. To understand an image, the CAD system breaks down the issue into multiple parts and does a step-by-step process to conclude whether a specific area on a radiology image looks suspicious. Therefore, it is important that radiologists have a basic understanding to comprehend why the output of the CAD is off from the usual even if it is because of human error.

10.1 Preprocessing

Most of the AI systems have an input or baseline dataset and starts preprocessing the data before it undergoes further changes through the scanning software. Series of calibrations are done to refine the data. These include resampling the data and removing the noises in the image. The basic reason for this process is to make sure that the existing dataset evolves. Since the AI system works on the knowledge from the previous data set which in turn comes from a binary data set (numbers 0 and 1) so the basic changes in the aberrancies from the data set can be easily pointed out for the doctors/radiologist to bring attention to. These aberrancies can be studied by the doctors (**Figure 2**) [3].

10.2 Demarcation

The second step is segmentation or demarcation of the normal structures in the data set. This step includes the demarcation, and consequent categorization of anatomic regions. This is the toughest step in the process and it is the most studied area. This also decides the accuracy of the AI system. As compared to the human knowledge which greatly relies on the differentiation of the structures and the artifacts from the prior knowledge the AI system greatly depend on the data sets. The more the data set the more refined algorithms are going to be (**Figure 3**) [28].

10.3 Detection of the aberrancy

The next big step is the aberrancy detected and identified matching the system to the subsets from the normal. These locations are called as the *candidates*. These *candidates* can be polyps, tumors, calcifications, dysplasia [3]. This step is very sensitivity driven. The AI system has to make sure the sensitivity crosses a certain threshold whereas the specificity can be low. It is important for us to understand these aberrancies may not necessarily be anomalies and are subject to the doctor's acumen to decipher and diagnose (**Figure 4**).

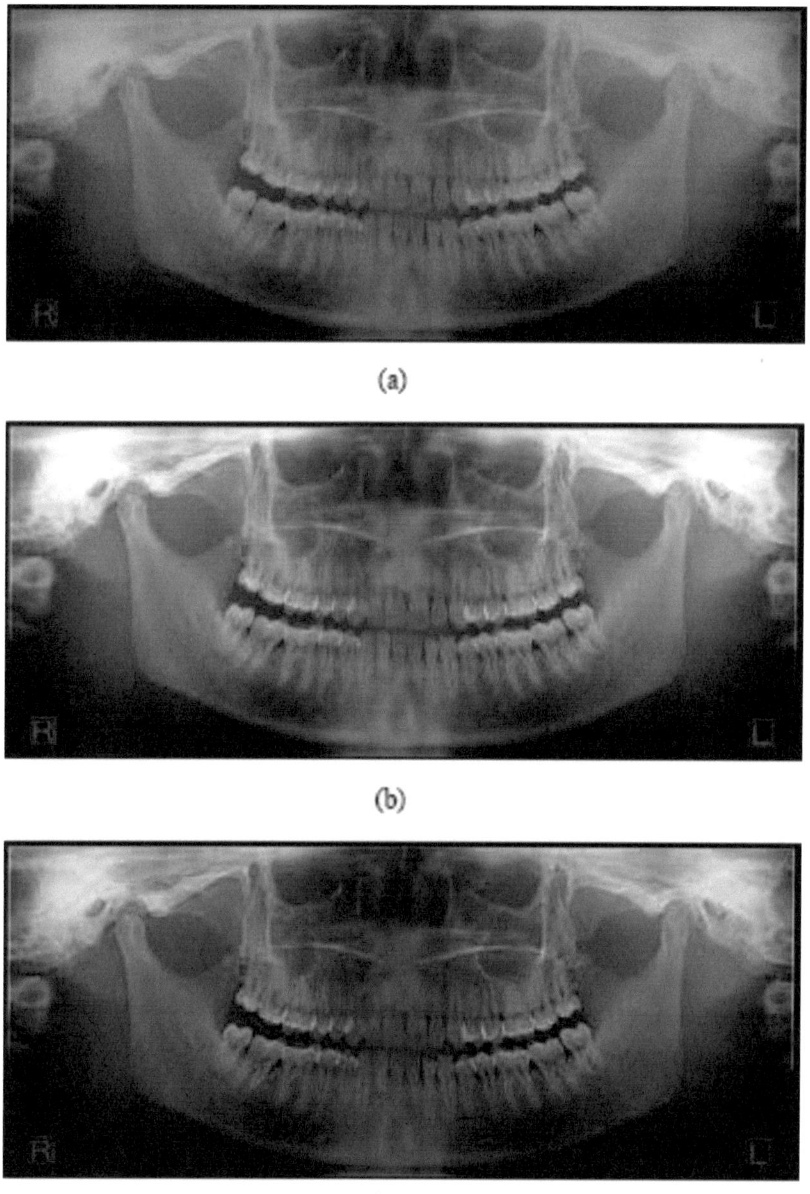

Figure 2.
(a) The CBCT reconstruction image with the crude image that was just captured from the patient and stored in to the system. (b) Enhanced image after the preprocessing with the color defects fixed. (c) Enhanced image after the preprocessing with the noise defects fixed.

10.4 Scrutinizing the aberrancy

As mentioned previously, the extracted data has to be very sensitive for the next stage, which is scrutinizing the aberrancy. In this step each area is closely analyzed to rule out the normal variations. This is done by using the *vector space paradigm*, which means each aberrancy from the normal is given a feature which can be computed. Each candidate has a feature and has its own mean value and standard deviations. The border gradients are also described accordingly. This is paramount

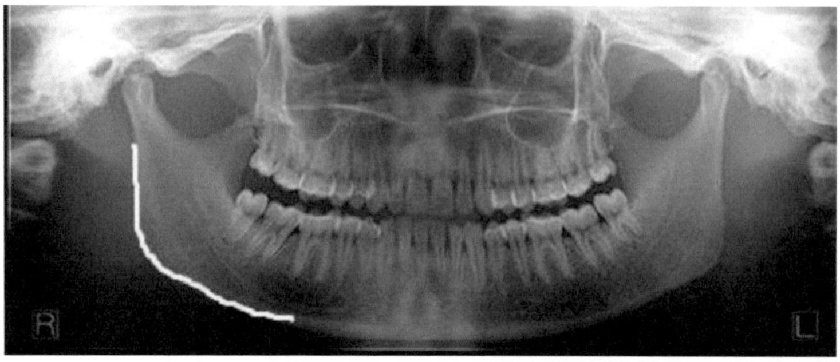

Figure 3.
CBCT panoramic reconstruction with demarcation of the anatomic structure showing the lower border of the mandible.

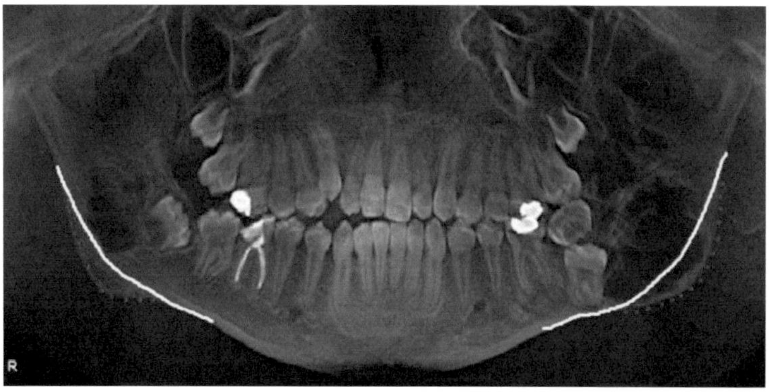

Figure 4.
CBCT panoramic reconstruction of a Cherubism showing aberrancy from the normal data set.

because now each aberrancy pointed out in the last step, is represented by a vector. These vectors can be mathematically represented in the space which is important to gauge anything in the machine's learning pattern, which is the basis of AI systems. These pattern analyses and machine learning can be used in any system to quantify the data and convert those into the computer's language [29–32].

10.5 Stratification

The learning patterns and the recognitions are classified and stratified in the space where the normal and the abnormal candidates exist. The next step is training the classified data set. Consistency is a big deciding factor here. The normal candidates are classified and stratified consistently with the in one subsets whereas the abnormal candidates are classified with consistency in the other data set. This provides training for the AI system to form an opinion regarding the classified and stratified data. This training is done with the help of person who had prior knowledge and could feed the data top provide reference standard with correct information. For example, the doctor can point out to the location of the cherubism on the CBCT pantographic reconstruction view with prior knowledge of its location. Also, data, like age, can be a deciding factor which is fed into the system by the doctor and this too aids the diagnostic process [29–32].

This is a very complicated step because some of the basic aberrancies which might not be the disease may also have the same locations or some diseases may not be in their classical locations. Therefore, data enrichment and stratification should be done on a regular basis.

The AI system, as discussed before, learns and adapts from the quality of the data set. So, we need more exposure to the data set to refine the results. AI researchers are always in the ongoing process of learning and experimenting with the classification schemes.

10.6 Output of AI

The process comes to completion when a degree of suspicion is assigned to each and every candidate in the strata or the group. The threshold decided by the doctor is the parameter used to test the degree of suspicion. These degrees of suspicion crossing the threshold are demarcated with the identifiers which can be circles or arrows. The AI system learns from the threshold and the final outcomes and enriches its data and subsets of the data. The results that come from the previous learning go to the computer's learning curve. For example, in the CBCT shown below if features of cherubism are detected, and the doctor diagnosed it as a case of cherubism, then the result is saved in the system for future reference for other cases. If CBCT for the case of a swollen angle of jaw presents a similar data, then the computer uses its previous knowledge to make sure whether the indicated region is considered true or false positive [3].

11. Clinical applications

There is a big range of applications of the AI and CAD in various fields of medicine. These applications have received a premarket approval (PMA). This encompasses the devices which have been shown to pose serious levels of risk for the users. In these cases the FDA guidelines recommend newer devices could be safer and more effective in the near future.

In the last decade there has been a considerable increase in their accuracy as systems for diagnostic help [25, 26]. The AI systems in mammography have been successful in detecting differences between the mass lesions and micro-calcifications [42]. For micro-calcifications it has shown that the AI systems show high performance and sensitivity which is greatly used by doctors in making an educated decision during diagnosis making. The data from multiple views and combination of different modalities like ultrasonography (US), magnetic resonance (MR) imaging, and digital breast tomosynthesis have shown to be really promising in enhancing the diagnosis and enriching the data for future diagnoses [34].

12. Computer simulation and the future of diagnostic expertise

In 1996, the American Journal of Roentgenology published a report of three cases of diagnostic errors in radiology. After assessing the clinicians' defense of their decisions, the author concluded that the radiologists missed out on the diagnosis because they did not think of the lesion rather than not know of it. It is popularly known as the *aunt Minnie effect*—that is if a woman in a picture looks like Aunt Minnie, she must be aunt Minnie [43]. Improving patient care means, we have to look for ways to minimize diagnostic errors- an umbrella term that includes as varied factors as personal or social bias, heuristics and even failure of perception [44]. Debiasing programs work but may take a lifetime to refine, even doctors who are willing to accept error on their part [44]. In the present world however, where software changes by the day, we do not happen to have that luxury of time.

Populations in countries have increased but investments in health and education have not kept pace with the same. Add to that an increasingly unstable world both politically and environmentally. All of these changes mean shifting populations and overburdened hospital based care provider. With the rise in longevity and polypharmacy, one diagnostician may not be and indeed, cannot be held accountable, for missing out on a few pertinent points here and there, while assessing a case. In today's outpatient medicine practice, especially, that of Oral Medicine and Radiology, there is hardly any time to think. And much to our chagrin, the reduced pay of practicing doctors, and increased work time of a hospital based doctor only means, what the New England Journal of Medicine, recently referred to as *subsistence-level intellectual mode* [45]. In such a scenario, the probability of diagnostic errors grows by leaps and bounds.

Where computer simulation steps in, is this very ripe environment of piling information and very few humans, qualified to process it. As of now, there are gaping holes in the way AI processes the information it collects, and human mind is, as yet, ahead of it. AI is heavily reliant on data sets that its algorithms work on. The stark problem with these data sets is that they are not representative of the wide gamut of humanity out there that needs treatment. For AI to pick a patient on its scanner, first of all the patient should have access to a device that lets this patient connect to its virtual world. Disadvantaged populations or displaced populations may not have that. This inherent bias of the system leaves it firmly in the field of speculation over its accuracy and therefore dependability.

Algorithms, however, tend to improve themselves over time; deleting redundancies and communicating across other platforms as they pick more and more data, as has been discussed earlier in the chapter. With the advent of quantum computing, in what has been defined by The Economist as the field's Sputnik moment, Google has recently demonstrated its ability to perform a task in little over 3 min, what might take most powerful classical computers about 10,000 years to complete [46]. In a world, where Common Health of Android and Apple Health of iOS, will increasingly guide the logistics of collecting and collimating data, medical or otherwise, from Electronic health records, medical devices, software, apps, and smartphones, the art of diagnosis will witness seismic changes. The quality of information being used to create data sets is one of the many hurdles.

Yet, as the horse carriage eventually gave way to sophisticated cars, we will have to yield a part of the field to computer simulation. But as we do that, we have to remember, the algorithm, as yet, does not have a totalitarian power over the mind. At the end of the day, the most important part of patient care is after all, care. And a responsive warm doctor in the field of Oral Medicine and Radiology is still very much preferred over subservience to any cold computing device. Computer simulations, then, are just another set of tools in our armamentarium and we need to research how to use the same for the benefit and better experience of everyone involved.

13. Conclusion

In a world, increasingly driven on ever evolving computation, diagnostic medicine has to adapt to change. Harnessing the power of AI to prevent logjams of channeling information collected into a cohesive whole will benefit both the doctor providing care and the patient receiving it. Computer simulation is a vital tool and how, and how much will it change the topography of diagnostic medicine, remains to be seen.

Author details

Saman Ishrat[1*], Akhilanand Chaurasia[2] and Mohammad Husain Khan[3]

1 Rama Dental College, Kanpur, India

2 King George's Medical University, Lucknow, India

3 Bidgley Inc., Mountain View, California, USA

*Address all correspondence to: samanishratalam@gmail.com

IntechOpen

© 2020 The Author(s). Licensee IntechOpen. This chapter is distributed under the terms of the Creative Commons Attribution License (http://creativecommons.org/licenses/by/3.0), which permits unrestricted use, distribution, and reproduction in any medium, provided the original work is properly cited.

References

[1] Miller RA. Why the standard view is standard: People, not machines, understand patients' problems. Journal of Medicine and Philosophy. 1990;**15**:581-591

[2] Miller RA. Computer-assisted diagnostic decision support: History, challenges, and possible paths forward. Advances in Health Sciences Education. 2009;**14**:89-106

[3] Ginneken BV, Schaefer-Prokop CM, Prokop M. Computer-aided diagnosis: How to move from the laboratory to the clinic. State of the Art. **261**(3):719-732

[4] Rubin GD. Data explosion: The challenge of multidetector-row CT. European Journal of Radiology. 2000;**36**(2):74-80

[5] Moore's Law. http://en.wikipedia.org/wiki/Moore's_law [Accessed: 26 August 2009]

[6] Kurzweil R. The Singularity Is near: When Humans Transcend Biology. New York, NY: Viking Penguin; 2005

[7] Ishida M, Kato H, Doi K, Frank PH. Development of a new digital radiographic image processing system. Proceedings of SPIE. 1982;**347**:42-48

[8] Ishida M, Frank PH, Doi K, Lehr JL. High-quality digital radiographic images: Improved detection of low-contrast objects and preliminary clinical studies. Radiographics. 1983;**3**:325-338

[9] Ishida M, Doi K, Loo LN, Metz CE, Lehr JL. Digital image processing: Effect on the detectability of simulated low-contrast radiographic patterns. Radiology. 1984;**150**:569-575

[10] Giger ML, Doi K. Investigation of basic imaging properties in digital radiography. Part 1: Modulation transfer function. Medical Physics. 1984;**11**:287-295

[11] Giger ML, Doi K. Investigation of basic imaging properties in digital radiography. Part 2: Noise wiener spectrum. Medical Physics. 1984;**11**:797-805

[12] MacMahon H, Vyborny CJ, Powell G, Doi K, Metz CE. The effect of pixel size on the detection rate of early pulmonary sarcoidosis in digital chest radiographic systems. Proceedings of SPIE. 1984;**486**:14-20

[13] Giger ML, Doi K. Investigation of basic imaging properties in digital radiography. Effect of pixel size on SNR and threshold contrast. Medical Physics. 1985;**12**:201-208

[14] Loo LN, Doi K, Metz CE. Investigation of basic imaging properties in digital radiography. 4. Effect of unsharp masking on the detectability of simple patterns. Medical Physics. 1985;**12**:209-214

[15] Fujita H, Doi K, Giger ML. Investigation of basic imaging properties in digital radiography. 6. MTFs of I.I.-TV digital imaging systems. Medical Physics. 1985;**12**:712-729

[16] MacMahon H, Vyborny CJ, Sabeti V, Metz CE, Doi K. The effect of digital unsharp masking on the detectability of interstitial infiltrates and pneumothoraces. Proceedings of SPIE. 1985;**555**:246-252

[17] Fujita H, Doi K, Giger ML, Chan HP. Investigation of basic imaging properties in digital radiograph. Characteristic curves of I.I.-TV digital systems. Medical Physics. 1986;**13**:13-18

[18] Giger ML, Doi K, Fujita H. Investigation of basic imaging properties in digital radiography. 7.

Noise wiener spectra of I.I.-TV digital imaging systems. Medical Physics. 1986;**13**:131-138

[19] MacMahon H, Vyborny CJ, Metz CE, Doi K, Sabeti V, Solomon SL. Digital radiography of subtle pulmonary abnormalities: An ROC study of the effect of pixel size on observer performance. Radiology. 1986;**158**:21-26

[20] Ohara K, Chan HP, Doi K, Fujita H, Giger ML. Investigation of basic imaging properties in digital radiography. 8. Detection of simulated low-contrast objects in DSA images. Medical Physics. 1986;**13**:304-311

[21] Giger ML, Ohara K, Doi K. Investigation of basic imaging properties in digital radiography. 9. Effect of displayed grey levels on signal detection. Medical Physics. 1986;**13**:312-318

[22] Kume Y, Doi K, Ohara K, Giger ML. Investigation of basic imaging properties in digital radiography. 10. Structure mottle of I.I.-TV digital imaging systems. Medical Physics. 1986;**13**:843-849

[23] Huang HK. PACS and Imaging Informatics: Basic Principles and Applications. John Wiley & Sons Inc.;

[24] Dean JC, Ilvento CC. Improved cancer detection using computer-aided detection with diagnostic and screening

[25] Brem RF. Clinical versus research approach to breast cancer detection with CAD: Where are we now? American Journal of Roentgenology. 2007;**188**(1):234-235

[26] Noble M, Bruening W, Uhl S, Schoelles K. Computer-aided detection mammography for breast cancer screening: Systematic review and meta-analysis. Archives of Gynecology and Obstetrics. 2009;**279**(6):881-890

[27] De Boo DW, Prokop M, Uffmann M, van Ginneken B, Schaefer-Prokop CM. Computer-aided detection (CAD) of lung nodules and small tumours on chest radiographs. European Journal of Radiology. 2009;**72**(2):218-225

[28] Jain AK. Fundamentals of Digital Image Processing. Upper Saddle River, NJ: Prentice Hall; 1989

[29] Duda RO, Hart PE, Stork DG. Pattern Classification. 2nd ed. New York, NY: Wiley; 2001

[30] Bishop CM. Pattern Recognition and Machine Learning. Secaucus, NJ: Springer; 2007

[31] Duin RP. Superlearning and neural network magic. Pattern Recognition Letters. 1994;**15**(3):215-217

[32] Bennett KP. Support vector machines: Hype or hallelujah? SIGKDD Explorations Newsletter. 2000;**2**(2):1-13

[33] van Ginneken B, Hogeweg L, Prokop M. Computer-aided diagnosis in chest radiography: Beyond nodules. European Journal of Radiology. 2009;**72**(2):226-230

[34] Giger ML, Chan HP, Boone J. Anniversary paper: History and status of CAD and quantitative image analysis: The role of medical physics and AAPM. Medical Physics. 2008;**35**(12):5799-5820

[35] Doi K. Computer-aided diagnosis in medical imaging: Historical review, current status and future potential. Computerized Medical Imaging and Graphics. 2007;**31**(4-5):198-211

[36] Nishikawa RM. Current status and future directions of computer-aided diagnosis in mammography. Computerized Medical Imaging and Graphics. 2007;**31**(4-5):224-235

[37] van Ginneken B, ter Haar Romeny BM, Viergever MA. Computer-aided diagnosis in chest radiography: A survey. IEEE Transactions on Medical Imaging. 2001;**20**(12):1228-1241

[38] Sluimer IC, Schilham AM, Prokop M, van Ginneken B. Computer analysis of computed tomography scans of the lung: A survey. IEEE Transactions on Medical Imaging. 2006;**25**(4):385-405

[39] Chan HP, Hadjiiski L, Zhou C, Sahiner B. Computer-aided diagnosis of lung cancer and pulmonary embolism in computed tomography—A review. Academic Radiology. 2008;**15**(5):535-555

[40] Boyer B, Balleyguier C, Granat O, Pharaboz C. CAD in questions/answers: Review of the literature. European Journal of Radiology. 2009;**69**(1):24-33

[41] Roehrig J. The manufacturer's perspective. The British Journal of Radiology. 2005;**78**(Spec No. 1):S41–S45

[42] Karssemeijer N, Bluekens AM, Beijerinck D, et al. Breast cancer screening results 5 years after introduction of digital mammography in a population-based screening program. Radiology. 2009;**253**(2):353-358

[43] Berlin L. Malpractice issues in radiology: Errors in judgment. American Journal of Roentgenology. 1996;**166**:1259-1261

[44] Croskerry P. The importance of cognitive errors in diagnosis and strategies to minimise them. Academic Medicine. 2003;**78**(8):775-780

[45] Ofri D. Perspective: Perchance to think. New England Journal of Medicine. 2019;**380**:1197-1199

[46] Quantum computing: Google claims to have demonstrated "quantum supremacy". The Economist. September 28, 2019

Chapter 6

Simulation and Parametric Inference of a Mixed Effects Model with Stochastic Differential Equations Using the Fokker-Planck Equation Solution

Bakrim Fadwa, Hamid El Maroufy and Hassan Ait Mousse

Abstract

This chapter is concerned with estimation method for multidimensional and nonlinear dynamical models including stochastic differential equations containing random effects (random parameters). This type of model has proved useful for describing continuous random processes, for distinguishing intra- and interindividual variability as well as for accounting for uncertainty in the dynamic model itself. Pharmacokinetic/pharmacodynamic modeling often involves repeated measurements on a series of experimental units, and random effects are incorporated into the model to simulate the individual behavior in the entire population. Unfortunately, the estimation of this kind of models could involve some difficulties, because in most cases, the transition density of the diffusion process given the random effects is not available. In this work, we focus on the approximation of the transition density of a such process in a closed form in order to obtain parameter estimates in this kind of model, using the Fokker-Planck equation and the Risken approximation. In addition, the chapter discusses a simulation study using Markov Chain Monte Carlo simulation, to provide results of the proposed methodology and to illustrate an application of mixed effects models with SDEs in the epidemiology using the minimal model describing glucose-insulin kinetics.

Keywords: stochastic differential equations, mixed effects model, Fokker-Planck equation, transition density, maximum likelihood estimators, genetic algorithm, Markov Chain Monte Carlo simulation

1. Introduction

In pharmacokinetic/pharmacodynamic studies, repeated measurements are performed on a sample of individuals (units/subjects), and responses for all experimental subjects are assumed to be described by a common structural model. This model contains both fixed and random effects to distinguish between individuals in a population, leading to a mixed effects model in which fixed effects represent fixed parameters for all individuals in the population and random effects account for individual differences. Moreover, mixed effects modeling has been shown to be useful in pharmacokinetic/pharmacodynamic studies; particularly for modeling total variation in and between individuals, the models used in pharmacokinetic/pharmacodynamic analysis are often presented by a system of deterministic differential equations (ordinary, partial, or delay). However, the pharmacological processes in reality are always exposed to incomprehensible effects that are difficult to model, and ignoring these effects can affect the parameter estimation results and their interpretations. So, the introduction of stochastic components to deterministic models is an important tool of analysis [1] and is more appropriate to model the intra-individual variations rather than ODEs. In addition, the extension of ODEs to the SDEs makes it possible to explain the differences between the observations and the predictions by two types of noise: dynamic noise that enters through the dynamics of the system and that can result from its random fluctuations or from the shortages of model and the measurement error which is added in the case of an indirectly observed process, which may be due to a test error or to the existence of a disturbance and represent the uncorrelated part of the residual variability. In the theory, there are rich and developed resources for mixed effects models whether deterministic [2–6] or stochastic and linear or nonlinear models (see many applications of stochastic NLME models in biomedical [7–9] and in pharmacokinetic [10–12] fields).

Parameter estimation in mixed effects models with SDEs, known by stochastic differential mixed effects models, is not an obvious procedure except in some cases simpler [14] because it is often difficult to write the likelihood function in its closed form. In this context, we propose a review on estimation methods of SDME models in [13, 15] and an example case that treats a generalized linear mixed models in [16] and also an example to approximate the likelihood function of an NLME with the likelihood of a linear mixed effects model in [17]. Moreover, to strengthen knowledge on estimation methods of SDME, we refer to [18, 19] that propose an example of stochastic mixed effects model with random effects log-normally distributed with a constant diffusion term.

In general, it is difficult to obtain an explicit likelihood function because the transition density of the stochastic process is often unknown or that the integral in the likelihood given the random effects cannot be computed analytically, and although the size of the random effects increases, the complexity of the problem increases also rapidly. Therefore, this requires a significant need for approximate methods to compute the transition density in an approximate closed form and for effective numerical integration methods to compute or approximate the integral in the likelihood function, using, for example, the Laplacian and Gaussian quadrature approximation [3, 8, 20] or other approaches [21, 22]. In the literature, several solutions have been proposed to approximate the transition density and have shown their effectiveness despite certain limitations. For example, the transition density could be approximated by the solution of the partial differential equations of Kolmogorov [23]; by the derivation of a Hermite expansion of closed form at the transition density [24–26] (this method has been reviewed and applied for many known stochastic processes for one-dimensional [8] and multidimensional [20] time-homogeneous SDME model); or by simulating the process to Monte-Carlo-integrate the transition density [27–29]. These techniques are very useful and can solve the problem, but unfortunately, they involve intense calculations which make the problem always complicated.

In this work, we focus on two fundamental issues concerning the implementation of SDEs in NLME models. The first is how the transition density of an SDME model can be approximated when it is not known, and the second is about approximating methods of the likelihood function when the integral given the random effects has no analytic solution. Then, we propose an optimization algorithm to obtain maximum likelihood estimators in order to facilitate the estimation procedure for these models. Finally, the methodology is evaluated by simulation studies on the bidimensional Ornstein-Uhlenbeck model, and then it is implemented on the minimal model describing the glucose and insulin kinetics.

2. Theoretical tools

Consider an N-dimensional continuous and stochastic process Y_t in the state space $E \subset \mathbb{R}^N$ described by the general first-order nonlinear stochastic differential equations of the Itô type [30]:

$$dY_t^i = \mu\left(Y_t^i, t, \theta, b^i\right)dt + \Sigma\left(Y_t^i, \theta, b^i\right)dW_t^i, \quad Y_0^i = y_0^i, \quad i = 1, \dots, M, \tag{1}$$

where Y_t^i is defined as the solution of the SDME model Eq. (1) that exists under some conditions that we suppose satisfied [31–33] and represents the observation of individual i from M different experimental units, $(i = 1, .., M)$, at the moment $t \geq t_0^i$, and $Y_0^i = Y_{t_0}^i$ is the initial state of Y_t for each subject. The process $\left\{\left(Y_t^i\right)_{t \geq 0}, i = 1..M\right\}$ is assumed to verify the same model structure Eq. (1) according to the individual parameters b^i; $\theta \in \Theta \subset \mathbb{R}^p$ is a p-dimensional fixed effects parameter which represents the same and common characteristics for all subjects, and $b^i \in B \subseteq \mathbb{R}^q$ are the q-dimensional individual random parameters assumed mutually independent, also called random effects because they are not the same for all individuals; they change between them according to a distribution of density $P_B\left(b^i|\Psi\right)$ depending on a population parameter Ψ; in the population approach, this parameter vector allows for a data from several subjects to be considered simultaneously. Each component b_l^i may follow a different distribution, $(l = 1, \dots, q)$, and a standard choice for the joint density function $P_B\left(b^i|\Psi\right)$ of the vector b^i could be the Gaussian distribution; however, any other distributions may be considered continuous or discrete:

$$b^i \sim i.i.d \;\; N(\vartheta, \phi). \tag{2}$$

The joint density function of the vector b^i is parameterized by a q-dimensional parameter $\vartheta \in \upsilon \subset \mathbb{R}^q$ and a $q \times q$-dimensional matrix $\phi \in \Phi \subset \mathbb{R}^{q \times q}$ representing the covariance matrix of b^i and specifying the parameters of the marginal distributions of the components b_l^i, $(1 \leq l \leq q)$; the components of ϕ and ϑ represent the population parameters Ψ. For Y_0^i, it is not necessarily known, and when its components are unknown, they must be considered as random effects since they change between individuals; but in some cases it can be known and assumed equal to a real constant. Also, we assume that the distribution of $Y^i(t)$ given $\left(b^i, \theta\right)$ and $Y^i(t') = y_{t'}, t' < t$, has a strictly positive density with regard to the Lebesgue measure on E:

$$y \rightarrow P_Y\left(y, t - t' | y_{t'}, b^i, \theta\right) > 0, \;\; y \in E. \tag{3}$$

$W^i(t)$ are the standard Brownian motions, and they are assumed mutually independent with b^j for all $1 \leq i, j \leq M$. The functions $\mu(\cdot) : E \times \mathbb{R} \times \Theta \times B \rightarrow \mathbb{R}$ and

$\Sigma(\cdot) : E \times \Theta \times B \rightarrow \mathbb{R}^+$ represent, respectively, the drift and the diffusion term of the model and are assumed to have some properties sufficiently regular to ensure a unique solution to the model [33].

According to the model Eq. (1), the process Y is the same and follows the same form for each individual of the population, and the SDME model of type Itô expresses the dynamic of the individual i perturbed by the Brownian motion. So, the differences between the subjects are modeled on the one hand by the different realizations of the Brownian motion paths $\{W_t^i\}_{t \geq t_0^i}$ and, on the other hand, by the incorporation of the random parameters b^i in the model. Therefore, the introduction of parameters varying randomly between subjects allows the model Eq. (1) to explain the variability between individuals.

The goal is to estimate the vector of fixed parameters θ and the parameter vector Ψ characterizing the distribution of random parameters b^is, but the statistical inference for such models is a difficult issue, and the level of difficulty is not the same whether the transition density is explicit or not and whether the process is observed directly or with measurement noise; in this work we assume that the process was exactly observed and no observation noise was considered.

2.1 Maximum likelihood estimation in SDME model

The likelihood function of an SDME model is expressed as follows:

$$L(\theta, \Psi) = \prod_{i=1}^{M} P\left(\underline{y}^i|\theta, \Psi\right) = \prod_{i=1}^{M} \int P_{\underline{Y}}\left(\underline{y}^i|b^i, \theta\right) P_B\left(b^i|\Psi\right) db^i \tag{4}$$

with

$$P_{\underline{Y}}\left(\underline{y}^i|b^i, \theta\right) = \prod_{j=1}^{n_i} P_Y\left(y_j^i, \Delta_j^i|y_{j-1}^i, b^i, \theta\right), \tag{5}$$

where n_i is the number of observations for the subject i at discrete points of time $\left\{t_0^i, t_1^i, \dots, t_{n_i}^i\right\}$, $i = 1, \dots, M$ and $\Delta_j^i = t_j^i - t_{j-1}^i$, $j = 1, \dots, n_i$. The conditional density $P_{\underline{Y}}\left(\underline{y}^i|\cdot\right)$ is equal to the product of the transition densities Eq. (5) for given random effects and θ, but the availability of the explicit transition density is the second constraint for the statistical issue of model Eq. (1) to obtain an exact likelihood function and exact estimators, since computing the transition density is not always obvious and requires approximation methods. However, there are some cases where the exact likelihood function is known, and the exact MLEs of θ are obtained (see references in the introduction). In fact, to compute the likelihood function in a closed form of an SDME model, we can encounter two types of problems that require approximate methods to overcome them: First, when the transition density $P_Y\left(y_j^i, \Delta_j^i|y_{j-1}^i, b^i, \theta\right)$ is known but the integral in Eq. (4) has no solution, in this case, the numerical methods of approximation of the integral are required. Or, second, when $P_Y\left(y_j^i, \Delta_j^i|y_{j-1}^i, b^i, \theta\right)$ cannot even be expressed explicitly and must also be approximated, see next paragraphs. Usually, in realistic examples, we have both an unknown transition density and an integral that is difficult to solve analytically. In theory, several methods for approximating transition densities and integrals have been proposed (see references cited in the introduction).

In this work, we propose an approximate method to estimate the transition density for a time-inhomogeneous NLME model with SDEs in a closed approximate form. So, we suggest to derive the transition density by solving the motion equations of the process using the Fokker-Planck equation, and we deal with the use of its solution given in [34]. Then, using the expression obtained, we get a closed form approximation of the likelihood function that we maximize to obtain the approximated MLEs $\hat{\theta}$ and $\hat{\Psi}$. The approximated transition density obtained from the proposed method is denoted by $P_Y^{(a)}\left(Y_j^i, \Delta_j^i | Y_{j-1}^i, b^i, \theta\right)$ that we substitute in Eq. (4) to obtain the following approximated likelihood function for the SDME model Eq. (1):

$$L^{(a)}(\theta, \Psi) = \prod_{i=1}^{M} \int \left(\prod_{j=1}^{n_i} P_Y^{(a)}\left(y_j^i, \Delta_j^i | y_{j-1}^i, b^i, \theta\right) \right) P_B\left(b^i | \Psi\right) db^i. \tag{6}$$

2.1.1 Laplace approximation

For a multidimensional vector of random parameters, if the exact transition density or its closed form approximation is available, we can use the Laplace approximation method [16, 20, 35] to obtain an explicit expression of the likelihood function to maximize, despite the fact that the integral in Eq. (4) has no closed solution. So, for a q-dimensional random vector b^i, the likelihood function Eq. (4) can be approximated as:

$$logL(\theta, \Psi) \simeq \sum_{i=1}^{M} \left[logP_{\underline{Y}}\left(\underline{y}^i | \tilde{b}^i, \theta\right) + logP_B\left(\tilde{b}^i | \Psi\right) + \frac{q}{2} log\,(2\pi) - \frac{1}{2} log\left| -H\left(\tilde{b}^i | \theta, \Psi\right)\right| \right], \tag{7}$$

where

$$\tilde{b}^i = argmax_{b^i}\left(f\left(b^i | \theta, \Psi\right)\right) \quad and \quad f\left(b^i | \theta, \Psi\right) = logP_{\underline{Y}}\left(\underline{y}^i | \tilde{b}^i, \theta\right) + logP_B\left(\tilde{b}^i | \Psi\right) \tag{8}$$

and $|\cdot|$ denotes the determinant of the Hessian matrix $H\left(b^i | \theta, \Psi\right)$:

$$H\left(\tilde{b}^i | \theta, \Psi\right) = \frac{\partial^2 \left[logP_{\underline{Y}}\left(\underline{y}^i | \tilde{b}^i, \theta\right) + logP_B\left(\tilde{b}^i | \Psi\right)\right]}{\partial \tilde{b}^i \partial \tilde{b}^{iT}} \tag{9}$$

2.2 Approximate transition density for multidimensional and nonlinear SDME model

Usually, a formal general solution of the stochastic differential equations as in Eq. (1) cannot be given, which makes the calculation of the transition density for this process more complicated. Moreover, this process has a lot of fluctuations that its exact position cannot be determined but can be known given a region by its probability density; with the FP equation, such a probability density can be determined. The FP equation is a differential equation for the distribution function describing a Brownian motion by which the probability density of the stochastic process can be calculated in a much simpler way by solving this equation. This motion equation is usually used for variables describing a macroscopic but small system, where the fluctuations are important as for some cases in physics, e.g., the position and the speed of the Brownian motion of a small particle. However, it can be also used for the larger system where, in spite of their small fluctuations, the stochastic description remains necessary when the deterministic equations may not be stable for this type of system.

So, under some assumptions, the transition density $P_Y\left(y_j^i, \Delta_j^i | y_{j-1}^i, b^i, \theta\right)$ of the process Y in the SDME model Eq. (1) is the solution of the following functional partial differential Equation [23, 34]:

$$\frac{\partial P_Y\left(y_j^i, \Delta_j^i | y_{j-1}^i, b^i, \theta\right)}{\partial t} = L_{FP} P_Y\left(y_j^i, \Delta_j^i | y_{j-1}^i, b^i, \theta\right), \tag{10}$$

where L_{FP} is the FP operator and $L_{FP} = -\sum_{k=1}^{N} \frac{\partial}{\partial Y_k}\left[\mu_k\left(Y^i, t, \theta, b^i\right)\right] + \frac{1}{2}\sum_{k=1}^{N}\sum_{l=1}^{N} \frac{\partial^2}{\partial Y_k \partial Y_l}\left[\Sigma_{kl}\right]$ and μ_k is the k-th element of the drift vector and Σ_{kl} is the kl-th element of the diffusion matrix. We assume that the diffusion matrix is positive definite so that its inverse exists $Det(\Sigma) \neq 0$.

In [34], H. Risken deals in this book with the derivation of the FP equation and its solution methods with some of its applications especially for problems of Brownian motion. So, here, we propose to characterize the transition density of the N-dimensional process Y_t^i in Eq. (1) using the Risken approximation based on the formal solution of Eq. (10) proposed in [34], p: 4.109, denoted by $P_Y^{(a)}\left(y_j^i, \Delta_j^i | y_{j-1}^i, b^i, \theta\right)$:

$$\begin{aligned} P_Y^{(a)}\left(Y_j^i, \Delta_j^i | Y_{j-1}^i, b^i, \theta\right) &= \left(2\sqrt{\Pi\Delta_j}\right)^{-N} [Det\Sigma]^{-\frac{1}{2}} * exp\left(-\frac{1}{4\Delta_j}\sum_{l=1}^{N}\sum_{k=1}^{N}\left[\Sigma^{-1}\right]_{lk}\right. \\ &\left.\left[\left(Y_j^i\right)_l - \left(Y_{j-1}^i\right)_l - \mu_l\left(Y_{j-1}^i, t, \theta, b^i\right)\Delta_j^i\right]\left[\left(Y_j^i\right)_k - \left(Y_{j-1}^i\right)_k - \mu_k\left(Y_{j-1}^i, t_{j-1}, \theta, b^i\right)\Delta_j^i\right]\right) \end{aligned} \tag{11}$$

To test the effectiveness of this approach, we will guide our statistical methodology by simulation studies in order to examine the flexibility of its application to deduce its advantages and disadvantages. So, we substitute $P_Y^{(a)}\left(Y_j^i, \Delta_j^i | Y_{j-1}^i, b^i, \theta\right)$ in Eq. (6), and by solving the integral with respect to the random effects density and maximizing the obtained likelihood function, we get the approximated estimators $\hat{\theta}$ and $\hat{\Psi}$.

2.2.1 Approximated estimators

For a nonlinear SDME model with Gaussian random effects using Eqs. (6), (7), and (11), we obtain the following approximated likelihood function:

$$\begin{aligned} logL^{(a)}(\theta, \Psi) \simeq \sum_{i=1}^{M}\Bigg[&-\frac{q}{2}log\left((2\pi)\right) - \frac{1}{2}log\left(det(\phi)\right) - \frac{n_i}{2}log\left(det(\Sigma)\right) + \left(\sum_{j=1}^{n_i} log\left(\left(2\sqrt{\Pi\Delta_j}\right)^{-N}\right) - \frac{1}{4\Delta_j}\right. \\ &\left.\left[\Sigma^{-1}\right]_{lk}\left[\left(Y_j^i\right)_l - \left(Y_{j-1}^i\right)_l - \mu_l\left(Y_{j-1}^i, t, \theta, \tilde{b}^i\right)\Delta_j^i\right]\left[\left(Y_j^i\right)_k - \left(Y_{j-1}^i\right)_k - \mu_k\left(Y_{j-1}^i, t_{j-1}, \theta, \tilde{b}^i\right)\Delta_j^i\right]\right) \\ &-\frac{1}{2}\left(\tilde{b}^i - \nu\right)'\phi^{-1}\left(\tilde{b}^i - \nu\right) + \frac{q}{2}log\left(2\pi\right) - \frac{1}{2}log\left(det\left(-H\left(\tilde{b}^i|\theta, \Psi\right)\right)\right)\Bigg]. \end{aligned} \tag{12}$$

The MLEs of (θ, Ψ) can be obtained using one of the optimization tools and numerical computation software, especially when it is complicated to compute the gradients of the likelihood analytically. Here, we propose to use the genetic algorithm as an optimization tool to maximize the approximate likelihood function Eq. (12) using MATLAB software:

$$\left(\hat{\theta}, \hat{\Psi}\right) = argmin_{GA}\left(-log\left(L^{(a)}(\theta, \Psi)\right)\right). \quad (13)$$

2.3 Genetic algorithm

The genetic algorithm is a random search technique to look for an exact or approximated optimum points for optimization problems [36–38]. It is based on the concepts of natural genetic evolution which contains the following stages: the reproduction, the crossing, and the mutation of a constantly evolving population. It sets up the evolution of a random population of potential solutions of the N cardinal; then, the N simultaneous iterative trajectories interact with each other by following or imitating the biological evolution, for a convergence of certain elements of the population towards an optimal point of the fitness function.

The GA can search in multiple directions to explore all the search space by the possibility of jumping across them, so that the seeds spread uniformly over the whole search space. In this algorithm, we have a diversity of initial populations which gives the global optimum faster than other algorithms, where the initial value is very important and should be enough close to the global optimum. All of these features allows the GA to be regarded as a driving tool of evolution giving good results for optimization processes [37, 39]. In the literature, there were many works about the application of GA in optimizing problems specially for likelihood function [40, 41]. For the use of GA, we must first define some parameters of the algorithm: Population size N, EN, SR, CP, MP, fitness function, and convergence criteria. In the following we present the GA steps:

Steps of GA:

1. Generate initial population $\left\{\beta_1^{(0)}, \beta_2^{(0)}, \dots, \beta_N^{(0)}\right\}, m = 0$ via an initialization strategy (random generation), in our case $\beta = (\theta, \Psi)$.

For $m = 0$:

2. Evaluate the Fitness function $log\left(-L^{(a)}\left(\theta^{(m)}, \Psi^{(m)}\right)\right)$.

3. While (convergence criteria are not satisfied):

Do:

4. Replacement step (by using SR and EN): At the SR rate, individuals with the worst results in step 2 of fitness function are replaced by new ones randomly generated, and a number EN of individuals is selected and accepted for the next step.

5. Selection operator by using roulette wheel method, based on the fact that the more the individual has a good result of fitness function, the more likely he will be selected.

6. Crossover operator by using CP and mutation operator by using MP: It is a mechanism of perturbation on the candidate individuals (parents) according to CP and MP to generate new groups of individuals, and we obtain a new $(m+1)nd$ population $\left\{\beta_1^{(m+1)}, \beta_2^{(m+1)}, \dots, \beta_N^{(m+1)}\right\}$.

Else:

7. Evolution stops; get GA output.

8. $m = m + 1$.

End For.

In the simulation studies paragraphs, the GA is implemented using the MATLAB software, where the function "ga," to generate the genetic algorithm, requires inputs that are chosen according to the constraints of each example (see the help window in MATLAB). Moreover, the algorithm parameters are chosen according to what is early used in the literature [39], $EN = 4,\ MP = 0.2,\ CP = 0.8\ and\ SR = 1/3$, and the search spaces are around the confidence interval of the minimal model parameters (see [42] and references therein).

3. Simulation studies for SDME model

3.1 The two-dimensional Ornstein-Uhlenbeck process

To apply the proposed methodology and evaluate its effectiveness, we consider the two-dimensional OU process that is very useful in pharmacokinetic/pharmacodynamic studies and in biology [43], physics, engineering, finance, and neuroscience applications [14, 44]. Indeed, the choice of this process is due to the fact that it is one of the few known multivariate SDME models with known transition density. For this reason, we choose the OU process to evaluate the methodology presented above, and we perform a comparison study between the results obtained using the proposed transition density in Eq. (11) and those obtained using the exact density. The model is defined as follows:

$$dY^{(1)i}(t) = -\left(\beta_{11}b^i_{11}\left(Y^{(1)i}(t) - \alpha_1\right) + \beta_{12}b^i_{12}\left(Y^{(2)i}(t) - \alpha_2\right)\right)dt + \Sigma_{11}dW^{(1)i}(t),\ \ Y^i_0 = y^{(1)i}_0, i = 1, \dots, M$$
$$dY^{(2)i}(t) = -\left(\beta_{21}b^i_{21}\left(Y^{(1)i}(t) - \alpha_1\right) + \beta_{22}b^i_{22}\left(Y^{(2)i}(t) - \alpha_2\right)\right)dt + \Sigma_{22}dW^{(2)i}(t),\ \ Y^i_0 = y^{(2)i}_0, i = 1, \dots, M \tag{14}$$

With $Y^i(t) = \begin{pmatrix} Y^{(1)i}(t) \\ Y^{(2)i}(t) \end{pmatrix}; \beta = \begin{pmatrix} \beta_{11} & \beta_{12} \\ \beta_{21} & \beta_{22} \end{pmatrix}; \alpha = \begin{pmatrix} \alpha_1 \\ \alpha_2 \end{pmatrix}; \Sigma = \begin{pmatrix} \Sigma_{11} & 0 \\ 0 & \Sigma_{22} \end{pmatrix};$ $W^i(t) = \begin{pmatrix} W^{(1)i}(t) \\ W^{(2)i}(t) \end{pmatrix}; Y^i(0) = \begin{pmatrix} Y^{(1)i}(0) \\ Y^{(1)i}(0) \end{pmatrix}$ and $b^i = \begin{pmatrix} b^i_{11} & b^i_{12} \\ b^i_{21} & b^i_{22} \end{pmatrix}$ where $b^i_{ll'}$ $i.i.d \sim \Gamma\left(r_{ll'}, r^{-1}_{ll'}\right),\quad l, l' = 1, 2;\ i = 1, \dots, M.$

We rewrite the system in matrix notation under the Itô formula; we denote by $(\cdot)$ the elementwise multiplication:

$$dY^i(t) = \beta \cdot b^i(\alpha - Y^i(t))dt + \Sigma dW^i(t),\ \ Y^i_0 = y^i_0,\ \ i = 1, \dots, M. \tag{15}$$

Here, the random effects b^i are a matrix and not a vector in order to have a uniform dimension writing of the Eq. (15) and are assumed mutually independent and independent of Y^i_0 and W^i. The fixed parameter vector is $\theta = (\beta_{11},\ \beta_{12},\ \beta_{21},\ \beta_{22}, \alpha_1,\ \alpha_2,\ \Sigma_{11},\ \Sigma_{22})$, and the population parameter vector is $\Psi = (r_{11},\ r_{12},\ r_{21},\ r_{22})$. The exact transition density of model Eq. (15) for a given realization of the random effects is a bivariate normal:

$$P_Y\left(Y^i_{t_j}, \Delta^i_j | Y^i_{t_{j-1}}, b^i, \theta\right) = (2\pi)^{-1}|\varsigma|^{\frac{-1}{2}} exp\left(\frac{-\left(Y^i_{t_j} - \mu\right)' \varsigma^{-1}\left(Y^i_{t_j} - \mu\right)}{2}\right), \tag{16}$$

with mean vector $\mu = \alpha + \left(Y^i_{t_{j-1}} - \alpha\right) exp\left(-(\beta.b^i)\Delta^i_j\right)$ and covariance matrix $\varsigma = \tau - \left(exp\left(-(\beta.b^i)\Delta^i_j\right)\ \tau\ exp\left(-(\beta.b^i)'\Delta^i_j\right)\right)$, where

$$\tau = \frac{1}{2tr(\beta.b^i)|\beta.b^i|}\left(|\beta.b^i|\Sigma\Sigma' + (\beta.b^i - tr(\beta.b^i).I)\Sigma\Sigma'(\beta.b^i - tr(\beta.b^i).I)'\right). \tag{17}$$

We assume that the matrices $\beta.b^i$ and Σ have full rank and the real parts of the eigenvalues of $\beta.b^i$ are positive definite in order that a stationary solution to Eq. (15) exists. Under these assumptions, we derive from Eqs. (14) and (11) the following approximated transition density of Y:

$$P^{(a)}_Y\left(Y^i_{t_j}, \Delta^i_j | Y^i_{t_{j-1}}, b^i, \theta\right) = \left(2\sqrt{\Pi\Delta_j}\right)^{-2}(\Sigma_{11}\Sigma_{22})^{-\frac{1}{2}} * exp\left(-\frac{1}{4\Delta_j}(\Sigma^{-1}_{11})\left[Y^{(1)i}_{t_j} - Y^{(1)i}_{t_{j-1}} + \left(\beta_{11}b^i_{11}\left(Y^{(1)i}_{t_{j-1}} - \alpha_1\right)\right.\right.\right.$$
$$\left.\left.\left. + \beta_{12}b^i_{12}\left(Y^{(2)i}_{t_{j-1}} - \alpha_2\right)\right)\Delta^i_j\right]^2 + (\Sigma^{-1}_{22})\left[Y^{(2)i}_{t_j} - Y^{(2)i}_{t_{j-1}} + \left(\beta_{21}b^i_{21}\left(Y^{(1)i}_{t_{j-1}} - \alpha_1\right) + \beta_{22}b^i_{22}\left(Y^{(2)i}_{t_{j-1}} - \alpha_2\right)\right)\Delta^i_j\right]^2\right). \tag{18}$$

In **Figure 1**, we illustrate in (a) the simulation of the OU process using the Euler scheme [45] with the following set of parameters:

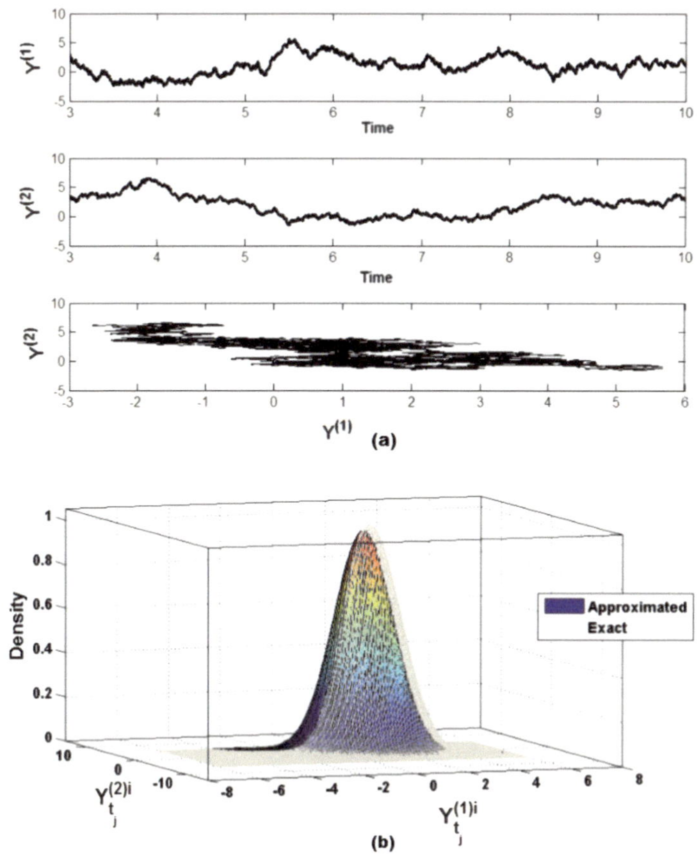

Figure 1.
A sample path of the OU process in the third graph of (a) for the given parameters set with the initial condition: $Y_0 = (3,3)$ and time interval [3,10]; and the transition density for a transition from Y_j to Y_{j+1} in (b).

$(\beta_{11} = 2.8, \beta_{12} = 2.5, \beta_{21} = 1.8, \beta_{22} = 2, \alpha_1 = 0.8, \alpha_2 = 1.5, \Sigma_{11} = 0.3, \Sigma_{22} = 0.5, r_{11} = 45, r_{12} = 100, r_{21} = 100,\ r_{22} = 125)$, and a time step of $\Delta_j = 0.001$ and represent in (b) the graph of the two transition densities given by Eqs. (16) and (18) for Y_j to Y_{j+1} using the same set of parameters and time step.

3.1.1 Simulation results

In this simulation study, we generate 1000 artificial datasets of dimension $2 \times (n+1) \times M$ from Eq. (15), where M is the number of subjects and n presents the number of repetitions of the experiment for each subject; then we estimate the parameters using the proposed approximated method, and we obtain 1000 sets of parameter estimates. The observations are obtained by linear interpolation from simulated trajectories using the Euler-Maruyama scheme with step size equal to 10^{-3}.

By plugging Eq. (16) in Eq. (4) and Eq. (18) in Eq. (6), we obtain a huge expression of the likelihood function but without a closed solution of integrals, so the exact estimators of θ and Ψ are unavailable. Therefore, in both cases, either using the exact or the approximated transition density, the Hessian matrix in Laplace approximation can be obtained analytically after a tedious calculation; then we apply the GA to obtain parameter estimates. But we cannot ignore the time consumed by this algorithm to get the results because of the long and complex expressions.

Table 1 shows that, for the given sample size, the results can be correctly identified using this estimated approach; even if some of parameters are overestimated or underestimated, the results remain acceptable because the results belong to the confidence interval. However, we believe that these results could be further proven by using other sample sizes and by adding alternative assumptions for the model that we did not consider in the methodology proposed in this chapter, which could further complicate the problem and be more time-consuming, as well as the present methodology suffers from some limitations. For the random parameters, the estimates can be provided using the optimization algorithm on Eq. (8) using the obtained estimate results of the parameters vector Ψ. Moreover, we conduct this simulation study using **Figure 2,** which shows that the empirical

True values				**Mean and (Std) (M = 40, n = 10)**			
β_{11}	β_{12}	β_{21}	β_{22}	$\hat{\beta}_{11}$	$\hat{\beta}_{12}$	$\hat{\beta}_{21}$	$\hat{\beta}_{22}$
2.8	2.5	1.8	2	3.10–3.25	2.48–2.56	1.72–1.65	2.11–2.37
				(0.164)–(0.283)	(0.015)–(0.283)	(0.095)–(0.189)	(0.153)–(0.255)
α_1	α_2	Σ_{11}	Σ_{22}	$\hat{\alpha}_1$	$\hat{\alpha}_2$	$\hat{\Sigma}_{11}$	$\hat{\Sigma}_{22}$
0.8	1.5	0.3	0.5	1.06–1.13	1.57–1.64	0.28–0.37	0.56–0.45
				(0.071)–(0.102)	(0.109)–(0.158)	(0.026)–(0.073)	(0.023)–(0.061)
r_{11}	r_{12}	r_{21}	r_{22}	$\hat{r}_{11}$	$\hat{r}_{12}$	$\hat{r}_{21}$	$\hat{r}_{22}$
45	100	100	25	44.75–52.43	100–112.75	102.35–89.64	24.72–31.02
				(0.523)–(9.372)	(01.166)–(28.113)	(01.04)–(22.05)	(2.297)–(11.20)

Table 1.
Ornstein-Uhlenbeck model: maximum likelihood estimates from 1000 simulations of model Eq. (14), using the exact and the approximated transition density.

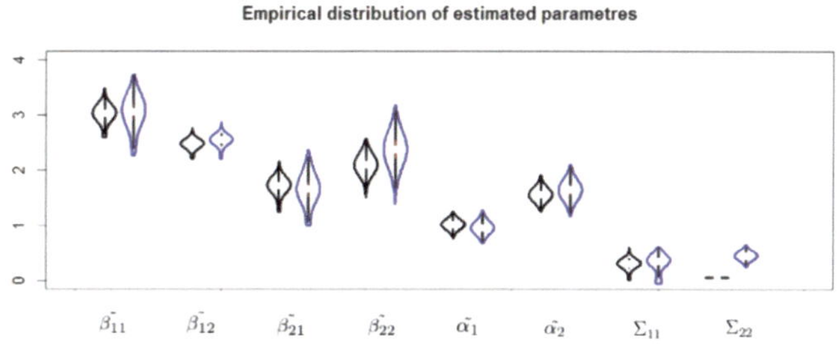

Figure 2.
Empirical distribution of estimates obtained using the exact and approximated transition density.

distribution of the most approximated estimators seems to be reasonably close to a normal distribution:

3.2 Stochastic minimal model

The minimal model describes the glucose-insulin kinetics and the dynamics of these processes illustrating the diabetes disease mechanisms. The diabetes is one of the most prevalent diseases in individuals and the nature and degree of its assignment changes from an individual to another and depends on certain individual characteristics, which implies that the concepts of stochastic modeling with random effects could be a good approach to modeling this disease. The Diabetes may be due to the insufficient insulin production (type 1 diabetes) or to the fact that the cells do not respond to the secreted insulin (type 2 diabetes) and the T2D patients tend to have substantially lower insulin sensitivity than healthy individuals, so that the T2D can be characterized by the level of insulin sensitivity for each individual. Therefore, to model the T2D, we observe how a person's body responds to insulin in the process of transporting glucose to tissues by measuring his insulin sensitivity. So, we present in this section the estimation of the minimal model which represents a powerful model describing the glucose-insulin kinetics for an individual's body in three differential equations, see the mathematical formulation of the model in [46–49]. So, it is already clear that the model will contain both fixed and random effects, because the study of diabetes disease takes into account the response of each individual according to his own parameters and other common parameters that describe the process of glucose-insulin for the entire population. See the description of the glucose-insulin kinetics in **Figure 3**.

From the mentioned literature and **Figure 3**, the glucose-insulin disposal can be represented, with respect to time, by the following nonlinear stochastic differential equations, perturbed by the stochastic terms $\sigma_1 dw_1(t)$, $\sigma_2 dw_2(t)$, and $\sigma_3 dw_3(t)$:

$$\begin{aligned} dG(t) &= \left[-\left(p_1 + X(t)\right)G(t) + p_1 G_b\right]dt + \sigma_1 dw_1(t), \;\; G(0) = G_0 \\ dX(t) &= \left[-p_2 X(t) + p_3 (I(t) - I_b)\right]dt + \sigma_2 dw_2(t), \;\; X(0) = 0 \\ dI(t) &= [-n(I(t) - I_b) + \gamma(G(t) - h)t]dt + \sigma_3 dw_3(t), \;\; I(0) = I_0, \end{aligned} \tag{19}$$

where $G(t)$ and $I(t)$ are, respectively, the concentration of glucose and insulin at time t in the blood. G_b and I_b indicate the basal level of glucose and insulin concentration before the glucose injection, this injection will cause a disturbance of the concentrations according to the mechanism described in these equations, and these values are assumed known for each individual. And, G_0 and I_0 are the theoretical measure of the concentrations at glucose injection moment at the beginning of the experiment.

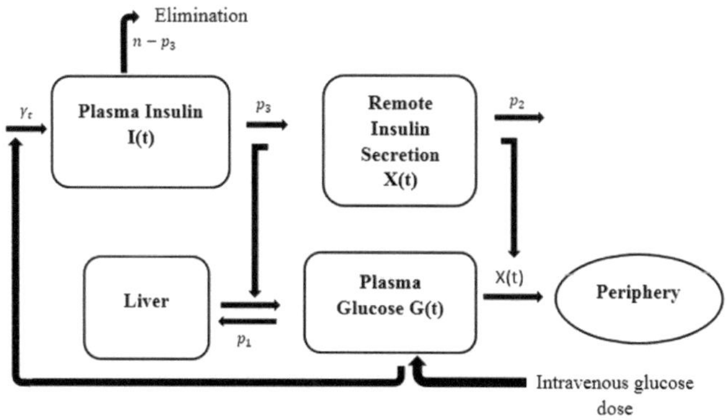

Figure 3.
At first, glucose and insulin concentrations in the blood are described by two sets of differential equations (see [50]); at a rate p_1, glucose leaves and enters the glucose space in proportion to the difference between the plasma glucose concentration G(t) and the basal plasma concentration G_b that represent the known pre-injection glucose level for each individual. Therefore, the parameter p_1 represents the glucose's own ability to be eliminated in muscles, liver, and tissues independently of insulin which is called glucose efficiency and denoted by S_G. Then, the glucose disappears from the glucose space at a rate proportional to insulin concentration in the insulin compartment X(t) which represents the dynamic of insulin response according to the two rates p_2 and p_3. These two parameters represent, respectively, the decreased glucose absorption capacity in tissues and its increased insulin dependency. For insulin secretion I(t), it is secreted by the pancreas independently of the glucose concentration, and proportional to a rate n to its own level already in the body and to the glucose level deferred from a threshold h at a rate γ when G(t) is above h, the insulin secretion does not only depend on the hyperglycemia level but also to the time spent since glucose injection.

The insulin sensitivity is defined by combining the two rates p_3 and p_2 as $S_I = p_3/p_2$, representing insulin's ability to increase the net glucose utilization [51].

Finally, the stochastic minimal model under the Itô sense re-parameterized by S_G and S_I can be rewritten as:

$$dY^i(t) = \begin{pmatrix} -\left(S_G^i + X(t)^i\right)G(t)^i + S_G^i G_b^i \\ -p_2\left(X(t)^i + S_I^i\left(I(t)^i - I_b^i\right)\right) \\ -n\left(I(t)^i - I_b^i\right) + \gamma\left(G(t)^i - h\right)t \end{pmatrix} dt + \Sigma dW(t); \quad Y_0^i = \begin{pmatrix} G_0^i \\ 0 \\ I_0^i \end{pmatrix}, \tag{20}$$

where $Y^i(t) = \begin{pmatrix} G(t)^i \\ X(t)^i \\ I(t)^i \end{pmatrix}$ and Σ is the diagonal diffusion matrix which contains the unknown constants σ_1, σ_2, and σ_3. The parameters $S_G^i, p_2, S_I^i, n, \gamma, h, G_0^i$ and I_0^i are unknown in the model and will be estimated. The parameters $S_G^i, S_I^i,\ I_0^i$ and G_0^i are assumed random, because they represent individual parameters that change from an individual to another. Each person has its own insulin sensitivity S_I^i which allows to know if the cells of his body react correctly or not to the insulin and if the insulin produced by the pancreas is sufficient or not, which can make some people with T2D and others without diabetes. Also, for glucose effectiveness S_G^i, which represents the glucose's own ability to be eliminated independently of insulin, it is unique to each individual and changes from a person to another, as well as for the measurement of glucose and insulin concentration. For the rest of the parameters, we consider them fixed since they describe the common side of the glucose-insulin process for the entire population (all individuals). So, we have the following random effect vector $b^i = \left(S_G^i,\ S_I^i,\ I_0^i,\ G_0^i\right)$, and we assume that:

$$S_G^i \sim N\left(\mu_{S_G}, \sigma_{S_G}\right),\ S_I^i \sim N\left(\mu_{S_I}, \sigma_{S_I}\right),\ I_0^i \sim N\left(\mu_{I_0}, \sigma_{I_0}\right),\ G_0^i \sim N\left(\mu_{G_0}, \sigma_{G_0}\right). \tag{21}$$

Random effects are assumed to be independent with a multinormal joint density function, with the mean $\vartheta = \left(\mu_{S_G}, \mu_{S_I}, \mu_{I_0}, \mu_{G_0}\right)$ and the covariance matrix $\phi = diag\left(\sigma_{S_G}, \sigma_{S_I}, \sigma_{I_0}, \sigma_{G_0}\right)$, so we have $\Psi = \left(\mu_{S_G}, \mu_{S_I}, \mu_{I_0}, \mu_{G_0}, \sigma_{S_G}, \sigma_{S_I}, \sigma_{I_0}, \sigma_{G_0}\right)$ and $\theta = (p2, n, \gamma, h, \sigma_1, \sigma_2, \sigma_3)$.

So, here we deal with a time-inhomogeneous NLME model with SDE describing the glucose-insulin kinetics (see [52] for the implementation of SDE time-inhomogeneous model) and [53] where the maximum likelihood estimation for a time-inhomogeneous stochastic differential model of glucose dynamics was treated. The measurements in the model Eq. (20) are assumed directly observed without measurement errors as well as in the theoretical approach presented above.

So, we wish to estimate (θ, Ψ) given the observations $\underline{y} = \left(\underline{y^1}, \ldots, \underline{y^M}\right)$ from model Eq. (20). By using the approximated transition density (Eq. (11)), we get the following approximated likelihood function for model Eq. (20):

$$\begin{aligned} L^{(a)}(\theta, \Psi) = \prod_{i=1}^{M} (2\pi)^{-2} (\sigma_{SG}\sigma_{SI}\sigma_{I0}\sigma_{G0})^{\frac{-1}{2}} \prod_{j=1}^{n_i} \left(\left(2\sqrt{\Pi\Delta_j}\right)^{-3}\right) [\sigma_1\sigma_2\sigma_3]^{-\frac{1}{2}}) \\ \int_{R^4} exp\left(\sum_{j=1}^{n_i}\left[\frac{-1}{4\Delta_j}\left[\frac{1}{\sigma_1}\left(A_{1j}^i\right)^2 + \frac{1}{\sigma_2}\left(A_{2j}^i\right)^2 + \frac{1}{\sigma_3}\left(A_{3j}^i\right)^2\right)\right]\right. \\ -\frac{1}{2}\left(\sigma_{S_G}^{-1}\left(S_G^i - \mu_{S_G}\right)^2 - \sigma_{S_I}^{-1}\left(S_I^i - \mu_{S_I}\right)^2 - \sigma_{I_0}^{-1}\left(I_0^i - \mu_{I_0}\right)^2 - \sigma_{G_0}^{-1}\left(G_0^i - \mu_{G_0}\right)^2\right)) dS_G{}^i dS_I{}^i dI_0{}^i dG_0{}^i \end{aligned} \tag{22}$$

with

$$A_{lj}^i = \left[\left(Y_j^i\right)_l - \left(Y_{j-1}^i\right)_l - \mu_l\left(Y_{j-1}^i, t_{j-1}, \theta, b^i\right)\Delta_j^i\right]. \tag{23}$$

We have no closed form solution to this integral, so exact estimators of θ and Ψ are unavailable. So, we use the Laplace approximation method described in Section 2 to obtain a closed form approximation to the log-likelihood function $log\left(L^{(a)}(\theta, \Psi)\right)$ for the model Eq. (20); then by applying the GA, we get the approximate estimates $\hat{\theta}$ and $\hat{\Psi}$.

3.2.1 Simulation results

We start this study with an application on artificial data generated on the intravenous glucose tolerance test principle (see [53] for a mathematical modeling of the test where glucose and insulin concentrations in plasma are subsequently sampled after an intravenous glucose injection). We generate 5000 sets of simulated artificial data of dimension $(n_i + 1) \times M$ from Eq. (20) using the Euler-Maruyama scheme [45] with a step size of 10^{-3} and a set of true parameters that are chosen according to [53, 54] representing the normal range of parameters values to simulate healthy subjects (without diabetes), with M being the number of units and n_i being the number of observations or repeated measurements collected on each unit i.

For each data from 5000 generated data sets, we estimate (θ, Ψ) by applying the proposed method. We first assume that the number of repeated measurements

True values				**(M,m)**	**Mean and (Std)**			
p_2	n	γ	h		$\hat{p}_2$	$\hat{n}$	$\hat{\gamma}$	$\hat{h}$
0.074	0.10	0.0007	90	(40,60)	0.0737 (0.0014)	0.100 (0.0013)	0.00073 (2.02×10^{-4})	89.08 (0.031)
				(40,10)	0.0784 (0.0059)	0.209 (0.0295)	0.00068 (2.54×10^{-4})	91.34 (0.712)
				(10,20)	0.0794 (0.0402)	0.156 (0.046)	0.00054 (0.0016)	62.11 (1.13)
σ_1	σ_2	σ_3	σ_{S_G}		$\hat{\sigma}_1$	$\hat{\sigma}_2$	$\hat{\sigma}_3$	$\hat{\sigma}_{S_G}$
0.01	0.06	0.03	0.006	(40,60)	0.009 (0.0012)	0.0616 (0.0011)	0.0343 (0.0027)	0.0061 (0.00010)
				(40,10)	0.014 (0.0023)	0.0708 (0.0060)	0.0340 (0.0035)	0.0073 (0.00030)
				(10,20)	0.015 (0.0031)	0.0938 (1.0196)	0.0480 (0.0108)	0.0067 (0.0006)
σ_{S_I}		σ_{I_0}	σ_{G_0}	μ_{S_G}	$\hat{\sigma}_{S_I}$	$\hat{\sigma}_{I_0}$	$\hat{\sigma}_{G_0}$	$\hat{\mu}_{S_G}$
0.000025	46	50	0.03	(40,60)	0.000021 (3×10^{-6})	44.98 (1.25)	45.96 (2.11)	0.0315 (0.0007)
				(40,10)	0.000029 (4.4×10^{-6})	43.94 (1.75)	45.16 (2.53)	0.0349 (0.0036)
				(10,20)	0.000016 (0.8×10^{-5})	41.09 (2.82)	44.64 (3.07)	0.0178 (0.0051)
μ_{S_I}	μ_{I_0}	μ_{G_0}			$\hat{\mu}_{S_I}$	$\hat{\mu}_{I_0}$	$\hat{\mu}_{G_0}$	
0.0002	95	320		(40,60)	0.00021 (1.2×10^{-6})	94.20 (1.12)	321.2 (1.07)	
				(40,10)	0.00025 (1.6×10^{-6})	92.05 (2.25)	318.6 (3.11)	
				(10,20)	0.00037 (1.35×10^{-5})	122.51 (2.51)	281.5 (4.88)	

Table 2.
Approximated maximum likelihood estimates and standard deviation from simulations of model Eq. (20), using the approximated transition density Eq. (11) with large and small DATA.

collected on each unit is constant $n_i = m$ and $\Delta_j^i = \Delta$ for all $1 \le i \le M$ and $1 \le j \le n_i$; then we apply the GA after choosing the good algorithm parameters (N, EN, SR, CP, MP), and we get 5000 sets of parameters estimates. We repeat this for large and small data with different possibilities of repetition of the experiment $(m, M) = (40, 60); (40, 10)$ and $(10, 20)$; for each parameter the sample mean and standard deviation are reported in **Table 2**. The simulation study on small data is treated in order to see if the size of the sample influences on the results and if the number of measures taken over time has a negligible effect or not, in other words, to see if it is possible to select only the essential measuring moments without repeating the measurements several times to well simulate a subject. We treat this issue in relation to our model and its study context, since in epidemiology the availability of data (measurements) at any point of time is an interesting constraint. We note that quantities of G_b and I_b are randomly simulated from the normal range of healthy subjects.

In **Table 2**, we report the results obtained on large and small data by maximizing Eq. (22) using the GA. We notice that, for the case of the large data, the true values

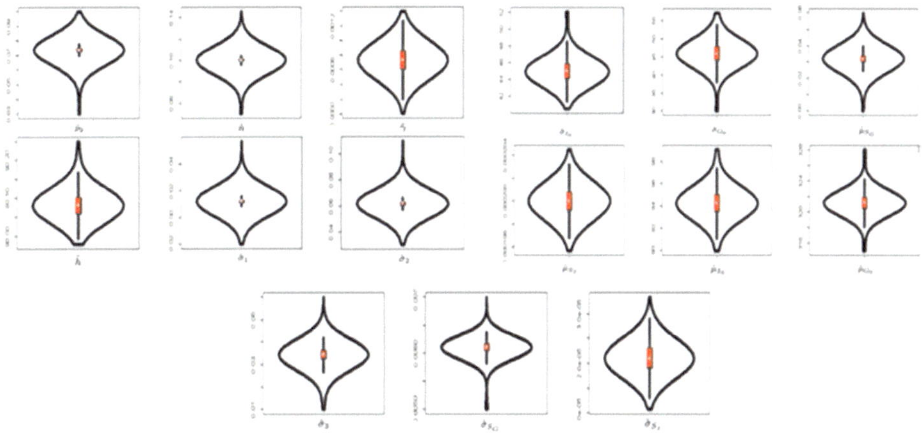

Figure 4.
SMM: empirical distribution of population parameter estimates obtained using (18) for (M, m) = (40,60).

of the parameters are well identified with the exception of some in which, in all the simulations and samples, the true value does not belong to the estimated confidence interval, such as σ_{S_I}, σ_{I_0}, and σ_{G_0}; nevertheless, the results are more satisfactory when the sample size M is large for all parameters. As mentioned before, we cannot ignore the limitations of this method, and the results could be improved by eliminating these limits and improving this approach, but for the given assumptions and tools, we can say that the results are still satisfactory even for a sample of $M = 10$ with 20 measures taken on each subject. So, we can conclude that we can rely on the results obtained from small samples with a small number of measurement repetitions of at least 20 observations. In addition, although we can use a relatively small number of measurements, we need to know how to choose the time points to perform the measurements in the blood, as this could, physiologically, affect selected observations and results. Thus, it is specified here that the essential task is to know how to choose the good moments of measurement after the injection, even in small numbers, chosen according to medical knowledge. **Figure 4** shows that, in the case of $(M; m) = (40; 60)$, the empirical distribution of the most approximated estimates seems to be reasonably close to a normal distribution.

Finally, from this simulation study where we have considered two SDME models, we can conclude that the parameters values of the models appear to be correctly identified using the proposed approach based on the Risken approximation to approximate the transition density of a stochastic process.

4. Conclusions

In this work, we have proposed a procedure to estimate the parameters of a mixed effects model containing stochastic differential equations, known by the SDME models, by proposing an approach to approximate its likelihood function to obtain the MLEs. This method has been evaluated by simulation studies on two SDME models in epidemiology: the two-dimensional Ornstein-Uhlenbeck process and stochastic minimal model. In fact, in models with SDEs instead of ODEs with random effects, the estimation of parameters is still not obvious even for one individual (one trajectory) because of the difficulties in deriving the transition densities, and difficulties become more interesting in using the population approach that treats the entire population simultaneously. The derivation of the exact density is not always possible for a stochastic and continuous process in an SDME model, so the search for an approximation of this density is an important step and requires an expensive calculation. This task is very interesting to give good results with good statistical properties of the estimators obtained by maximizing the likelihood function. In this work, the proposed estimation method has been applied to multidimensional and nonlinear SDME models with many random parameters normally distributed that can be extended to random parameters of any distribution. So, an approximation of the transition density $P_Y^{(a)}$ is obtained in a closed form using the Risken approximation for the formal solution of the Fokker-Planck equation proposed by Risken [34], and then the approximated likelihood is obtained using the Laplace approximation method and optimized using the genetic algorithm; these calculation procedures can be obtained using any numerical calculation software or with symbolic computing capabilities.

The classical inference of SDME models implies the problem of the numerical evaluation of the integral for the given random effects in the likelihood function, which becomes complicated especially when the model contains more than two random parameters. In the literature, several methods have been proposed and tested for the approximation of the integral (see references in the introduction) and the following examples: [8] which proposes the Gaussian quadrature method to solve the integrals for the case of an SDME models with a single random effect and [20] where the study was revised for a general case with several random parameters using the Laplace approximation to compute the integral in Eq. (4) and Eq. (6) numerically. For the mixed effects framework, see [3, 16, 55, 56]. In the case of using the Laplace method, as in this chapter, the calculation of the Hessian matrix can be done analytically when it is possible, as the examples in Section 2, or with the help of a symbolic calculus software or the automatic differentiation (AD) tools [57].

The results of simulation studies are satisfactory and can be obtained either by using moderate values for the number of experimental units M and of observations n taken for each experimental unit or by using a small sample size but with a number of measurements taken for each subject of 10 at least; this is relevant for applications where large sets of data are not available, such as biomedical applications where the mixed effect theory is widely applied.

The advantages of this approach, compared to those proposed in the literature for multidimensional SDME models with more than one random parameter [26], are that the computation of the approximate density is very easy and does not require a lot of time to calculate it or to program it in a software; the only task that can be time-consuming is in the optimization step to search for the optimum solution of the likelihood function, and also the proposed method is effective even with large data with a MATLAB program on a common PC (Intel Pentium IV 3.0 GHz with 512 MB of RAM). Nevertheless, the method suffers some limitations, for example, when the conditions to use Eq. (11) are not verified when, e.g., the inverse of the diffusion term does not exist and when, in certain cases, it is not obvious to derive the gradients and Hessians terms. Another limitation is that measurement errors are not considered in this work, and for a good stochastic version, it will be better to include noise on process increments and noise on observations that may be significant compared to system noise. These limitations may provide a perspective towards a more elaborate extension of the statistical study for SDME models, particularly in the field of epidemiology.

In conclusion, in this work, we proposed a method of parameter estimation for a mixed effects models with SDEs proposing an approximation method for the transition density in the case when it cannot be obtained in a closed form, with an approximation method of integral computation for the case of an SDME model with several random parameters. We have treated two examples to illustrate the effectiveness of this approach using computer tools. Indeed, we believe that this type of models is very interesting and provides a powerful and flexible modeling approach for repeated measurement studies such as biological and pharmacokinetic/pharmacodynamic and financial studies, as they combine the good characteristics of mixed effects and stochastic increments in intra-subject dynamics for a good modeling of a phenomenon.

Acknowledgements

Z.T., H.E., and F.B conceived the presented idea and contributed to the design and implementation of the research. F.B., H.E., Z.T., and H.A contributed to the analysis of the results. Both F.B. and H.E authors developed the theory and performed the computations and the numerical simulations and verified the analytic methods and calculations.

Conflict of interest

The author(s) declare no competing interests.

Abbreviations

SDEs	stochastic differential equations
MCMC	Markov Chain Monte Carlo
FP	Fokker-Planck equation
MLEs	maximum likelihood estimators
ODE	ordinary differential equations
NLME	nonlinear mixed effects
SDME	stochastic differential mixed effects
OU	Ornstein-Uhlenbeck
GA	genetic algorithm
EN	elite number
SR	selection rate
CP	crossover probability
MP	mutation probability
SMM	stochastic minimal model
T2D	type 2 diabetes

Author details

Bakrim Fadwa[1*†], Hamid El Maroufy[1†] and Hassan Ait Mousse[2]

1 Faculty of Science and Techniques, Department of Applied Mathematics, Sultan Moulay Slimane University, Beni-Mellal, Morocco

2 Faculty of Sciences and Techniques, Sultan Moulay Slimane University, Beni Mellal, Morocco

*Address all correspondence to: bakrim.fadwa@gmail.com

† These authors are contributed equally.

IntechOpen

© 2020 The Author(s). Licensee IntechOpen. This chapter is distributed under the terms of the Creative Commons Attribution License (http://creativecommons.org/licenses/by/3.0), which permits unrestricted use, distribution, and reproduction in any medium, provided the original work is properly cited.

References

[1] Allen E. Modeling with Itô Stochastic Differential Equations. Springer Science and Business Media, Springer; 2007

[2] Vonesh E, Chinchilli V. Linear and Nonlinear Models for the Analysis of Repeated Measurements. New York: Marcel Dekker; 1997

[3] McCulloch CE, Searle SR. Generalized, Linear and Mixed Models. New York: Wiley; 2001

[4] Kuhn E, Lavielle M. Maximum likelihood estimation in nonlinear mixed effects models. Computational Statistics & Data Analysis. 2005;**49**:1020-1038

[5] Guedj J, Thiébaut R, Commenges D. Maximum likelihood estimation in dynamical models of HIV. Biometrics. 2007;**63**:1198-1206

[6] Wang J. EM algorithms for nonlinear mixed effects models. Computational Statistics & Data Analysis. 2007;**51**: 3244-3256

[7] Picchini U, Ditlevsen S, De Gaetano A. Modeling the euglycemic hyperinsulinemic clamp by stochastic differential equations. Journal of Mathematical Biology. 2006;**53**:771-796

[8] Picchini U, De Gaetano A, Ditlevsen S. Stochastic differential mixed-effects models. Scandinavian Journal of Statistics. 2010;**37**:67-90

[9] Ditlevsen S, Yip K-P, Marsh D, Holstein-Rathlou N-H. Parameter estimation of the feedback gain in a stochastic model of renal hemodynamics: Differences between spontaneously hypertensive rats and Sprague-Dawley rats. American Journal of Physiology. Renal Physiology. 2007; **292**:607-616

[10] Sheiner L, Beal S. Evaluation of methods for estimating population pharmacokinetic parameters. I. Michaelis-Menten model: Routine clinical pharmacokinetic data. Journal of Pharmacokinetics and Biopharmaceutics. 1980;**8**:553-571

[11] Sheiner L, Beal S. Evaluation of methods for estimating population pharmacokinetic parameters. II. Biexponential model and experimental pharmacokinetic data. Journal of Pharmacokinetics and Biopharmaceutics. 1981;**9**:635-651

[12] Donnet S, Samson A. A review on estimation of stochastic differential equations for pharmacokinetic/ pharmacodynamic models. Advanced Drug Delivery Reviews. 2013;**65**(7): 929-939

[13] Donnet S, Samson A. Parametric inference for mixed models defined by stochastic differential equations. European Series in Applied and Industrial Mathematics. 2008;**12**: 196-218

[14] Ditlevsen S, De Gaetano A. Mixed effects in stochastic differential equations models. REVSTAT. 2005;**3**:137-153

[15] Donnet S, Samson A. Parametric inference for mixed models defined by stochastic differential equations. European Series in Applied and Industrial Mathematics. 2008;**12**: 196-218

[16] Pinheiro J, Chao E. Efficient Laplacian and adaptive Gaussian quadrature algorithms for multilevel generalized linear mixed models. Journal of Computational and Graphical Statistics. 2006;**15**:58-81

[17] Lindstrom M, Bates D. Nonlinear mixed-effects models for repeated measures data. Biometrics. 1990;**46**: 673-687

[18] Overgaard R, Jonsson N, Tornoe C, Madsen H. Nonlinear mixed-effects

models with stochastic differential equations: Implementation of an estimation algorithm. Journal of Pharmacokinetics and Pharmacodynamics. 2005;**32**:85-107

[19] Tornøe C, Overgaard R, Agerso H, Nielsen HA, Madsen H, Jonsson EN. Stochastic differential equations in NONMEM: Implementation, application, and comparison with ordinary differential equations. Pharmaceutical Research. 2005;**22**: 1247-1258

[20] Picchini U, Ditlevsen S. Practical estimation of high dimensional stochastic differential mixed-effects models. Computational Statistics & Data Analysis. 2011;**55**: 1426-1444

[21] Fröberg C. Numerical Mathematics: Theory and Computer Applications. Redwood City, CA: Benjamin-Cummings Publishing Company; 1985

[22] Krommer AR, Ueberhuber CW. Computational integration. Philadelphia, PA: Society for Industrial and Applied Mathematics; 1998

[23] Lo A. Maximum likelihood estimation of generalized Ito processes with discretely-sample data. Econometric Theory. 1988;**4**:231-247

[24] Aït-Sahalia Y. Comment on numerical techniques for maximum likelihood estimation of continuous-time diffusion processes by G. Durham and A. Gallant. Journal of Business & Economic Statistics. 2002;**20**:317-321

[25] Aït-Sahalia Y. Maximum likelihood estimation of discretely sampled diffusions: A closed-form approximation approach. Econometrica. 2002b;**70**:223-262

[26] Aït-Sahalia Y. Closed-form likelihood expansion for multivariate diffusions. Ann. Statist. 2008;**36**: 906-937

[27] Nicolau J. A new technique for simulating the likelihood of stochastic differential equations. The Econometrics Journal. 2002;**5**:91-103

[28] Hurn A, Lindsay K, Martin V. On the efficacy of simulated maximum likelihood for estimating the parameters of stochastic differential equations. Journal of Time Series Analysis. 2003; **24**:45-63

[29] Ripley B. Stochastic Simulation (Paperback Edition, 2006). New York: Wiley; 1987

[30] Ramanathan M. An application of ito's lemma in population pharmacokinetics and pharmacodynamics. Pharmaceutical Research. 1999;**16**:584-586

[31] Andersson H, Britton T. Stochastic Epidemic Models and their Statistical Analysis. New York: Springer Lecture Notes in Statistics; 2000

[32] Becker NG. On a general stochastic epidemic model. Theoretical Population Biology. 1976;**11**:23-26

[33] Øksendal B. Stochastic Differential Equations: An Introduction with Applications. 6th ed. Berlin-Heidelberg: Springer-Verlag; 2007

[34] Risken H. The Fokker-Plank Equation, Method of Solution and Application. 2nd ed. New York: Springer-Verlag; 1989

[35] Shun Z, McCullagh P. Laplace approximation of high dimensional integrals. Journal of the Royal Statistical Society B. 1995;**57**(4):749-760

[36] Goldberg DE. Genetic Algorithms in Search. Addison-Wesley Reading Mass: Optimization and Machine Learning; 1989

[37] Mickalewicz Z. Genetic Algorithms + Data Structure = Evolution Programs. Artificial Intelligence, Berlin: Springer-Verlag; 1992

[38] Sivanandam SN, Deepa SN. Introduction to Genetic Algorithms. Berlin: Springer-Verlag; 2008

[39] Jong D, Kenneth A. Evolutionary Computation: A Unified Approach. Cambridge, MA, United States: MIT Press; 2006

[40] Yalçinkaya A, Şenoglŭ B, Yolcu U. Maximum likelihood estimation for the parameters of skew normal distribution using genetic algorithm. Swarm and Evolutionary Computation. 2017;**38**: 127-138. DOI: 10.1016/j.swevo.2017.07.007

[41] Petrovski A, Wilson A, McCall J. Statistical Analysis of Genetic Algorithms and Inference about Optimal Factors. SCMS Technical Report 1998/2. Aberdeen, Scotland: The Robert Gordon University; 1998

[42] Andersen KE, Malene H. A population-based Bayesian approach to the minimal model of glucose and insulin homeostasis. Statistics in Medicine. 2005;**24**:2381-2400

[43] Favetto B, Samson A. Parameter estimation for a bidimensional partially observed OrnsteinUhlenbeck process with biological application. Scandinavian Journal of Statistics. 2010; **37**:200-220

[44] Picchini U, Ditlevsen S, De Gaetano A. Maximum likelihood estimation of a time-inhomogeneous stochastic differential model of glucose dynamics. Mathematical Medicine and Biology: A Journal of the IMA. 2008; **25**(2):141-155

[45] Kloeden P, Platen E. Numerical Solution of Stochastic Differential Equations. Berlin-Heidelberg: Springer-Verlag; 1992

[46] Bergman RN, Bortolan G, Cobelli C, Toffolo G. Identification of a minimal model of glucose disappearance for estimating insulin sensitivity. Proceedings of IFAC Symposium Identification and System Parameter Estimation. 1979;**2**:883-890

[47] Bergman RN, Phillips LS, Cobelli C. Physiologic evaluation of factors controlling glucose tolerance in man. Journal of Clinical Investigation. 1981; **68**:1456-1467

[48] Toffolo G, De Grandi F, Cobelli C. Estimation of beta-cell sensitivity from intravenous glucose tolerance test C-peptide data. Knowledge of the kinetics avoids errors in modeling the secretion. Diabetes. 1995;**44**:845-854

[49] Cobelli C, Man CD, Sparacino G, Magni L, De Nicolao G, Boris Kovatchev P. Diabetes: Models, signals, and control. IEEE Reviews in Biomedical Engineering. 2009;**2**:54-96

[50] Pacini G, Bergman RN. MINMOD: A computer program to calculate insulin sensitivity and pancreatic responsivity from the frequently sampled intravenous glucose tolerance test. Computer Methods and Programs in Biomedicine. 1986;**23**:113-122

[51] Bergman RN, Ider YZ, Bowden CR, Cobelli C. Quantitative estimation of insulin sensitivity. American Journal of Physiology-Endocrinology And Metabolism. 1979;**236**(6):E667

[52] Egorov A, Li H, Xu Y. Maximum likelihood estimation of time-inhomogeneous diffusions. Journal of Econometrics. 2003;**114**:107-139

[53] Gaetano A, Arino O. Mathematical modelling of the intravenous glucose tolerance test. Journal of Mathematical Biology. 2000;**40**:136-168

[54] Makroglou A, Li J, Kuang Y. Mathematical models and software tools for the glucose-insulin regulatory system and diabetes: An overview. Applied Numerical Mathematics. 2006; **56**(3):559-573

[55] Davidian M, Giltinan D. Nonlinear models for repeated measurements: An overview and update. Journal of Agricultural, Biological, and Environmental Statistics. 2003;**8**: 387-419

[56] Pinheiro J, Bates D. Approximations of the log-likelihood function in the nonlinear mixedeffects model. Journal of Computational and Graphical Statistics. 1995;**4**:12-35

[57] Griewank A. Evaluating Derivatives: Principles and Techniques of Algorithmic Differentiation. Philadelphia, PA: SIAM; 2000

Chapter 7

The Possibilities of Modeling Petri Nets and Their Extensions

Goharik Petrosyan

Abstract

This chapter is dedicated to several structure features of Petri nets. There is detailed description of appropriate access in Petri nets and reachable tree mechanism construction. There is an algorithm that describes the minimum sequence of possible transitions. The algorithm developed by us finds the shortest possible sequence for the network promotion state, which transfers the mentioned network state to the coverage state. The corresponding theorem is proven, which states that due to the describing algorithm, the number of transitions in the covering state is minimal. This chapter studies the interrelation of languages of colored Petri nets and traditional formal languages. The Venn diagram, modified by the author, is presented, which shows the relationship between the languages of the colored Petri nets and some traditional languages. As a result, it is shown that the language class of colored Petri nets includes an entire class of context-free languages and some other classes. The results obtained show that it is not possible to model the Patil problem using the well-known semaphores P and V or classical Petri nets, so the mentioned systems have limited properties.

Keywords: petri nets, colored petri nets, traditional languages, transition, position

1. Introduction

Modeling and designing systems cannot be imagined without the use of computer technology. When creating automated systems and designing them, the problem of choosing a formal model for representing systems first arises. From the model through the algorithmic to the software—this is the way of modern modeling and system design. When considering lumped physical systems, a convenient model is a linear graph, each vertex of which corresponds to a functional or constructive component, and an arc to a causal relationship.

Petri nets are a mathematical apparatus for modeling dynamic discrete systems. Their feature is the ability to display parallelism, asynchrony, and hierarchy. They were first described by Karl Petri in 1962.

The Petri net is a bipartite oriented graph consisting of two types of vertices—positions and transitions—connected by arcs between each other; vertices of the same type cannot be directly connected. Positions can be placed tags (markers) that can move around the network [1].

Petri net—a tool for modeling dynamic systems. The theory of Petri nets makes it possible to model a system with a mathematical representation of it in the form of

IntechOpen

a Petri net, the analysis of which helps to obtain important information about the structure and dynamic behavior of the simulated system.

There are several ways of practical application of Petri nets in the design and analysis of systems. In one of the approaches, the Petri nets are considered as an auxiliary analysis tool. Here, to build the system, generally accepted design methods are used, then the constructed system is modeled by the Petri net, and the constructed model is analyzed.

In another approach, the entire process of design and characterization is carried out in terms of Petri nets. In this case, the task is to transform the representation of the Petri net into a real information system [2].

The undoubted advantage of Petri nets is a mathematically rigorous description of the model. This allows their analysis with the help of modern computing techniques (including those with a massively parallel architecture) [1].

In modern society, reliable transmission and protection of information are of wide use and are topical tasks. The main task of Petri nets is the modeling of realistic systems from the point of view of optimization. Systematic study of the properties of Petri nets and the possibility of using them for solving applied problems, mainly problems related to models and means of parallel processing of information.

The following issues can serve as examples of those problems that often arise in the design and study of discrete systems:

- Does the system perform the functions for which it is intended?
- Does it function effectively?
- Can mistakes and emergencies occur in it?
- Does it have potential bottlenecks?
- Is it possible to simplify the system or replace its individual components and subsystems with more perfect ones, without disturbing its overall functioning?
- Is it possible to design more complex systems that meet the specified requirements from these systems, etc.?

These tasks are basically "qualitative" not quantitative.

The goal of in-depth study of various extensions of Petri nets (from the point of view of optimization) for modeling real-time systems brings to the design of such technical equipment where one has to minimize resource costs and time and maximize speed.

Colored petri net (CPN) modeling mechanisms are a convenient graphic language for designing, modeling, and testing systems [3–7]. They are well suited for systems that discuss interaction issues and synchronize. The colored Petri nets are well suited for modeling distributed systems, automated production systems, and for the design of VLSI circuit chips [8–10].

Colored Petri nets are called if the chips are the values of some types of data, which are usually called color sets. Expressions are assigned to arcs in such a network. When transitions are triggered, the values of expressions on arcs are calculated. The results of the calculations are extracted from the markup of the input transition points and placed in the marking of the output points. Transitions may be assigned with security expressions. If the guard expression assumes the value "false," the transition is prohibited [3–6, 11, 12].

The language generated by CPN allows to represent a model that is a collection of modules, allowing you to hierarchically represent complex networks or systems.

In classical Petri nets, the tokens do not differ from each other; they are colorless. In colored Petri nets, a position can contain chips that are of arbitrary complexity, such as lists, etc., that allow you to simulate more reliable models [8–10, 13].

2. The algorithm description of the shortest possible sequence of transitions in petri nets

To build models of discrete systems, it needs various components of the system with abstract operations: switching the transition from one state to another; the action of a program operator, machine, or conveyor; interruptions in the operating system; phase completion in the project; etc. The same system can work differently under different conditions, generating many processes that will bring nondeterministic work. In real systems, cases occur at certain periods and last a certain time. In synchronous models of discrete systems, events are correctly associated with certain pauses, moments during which all components simultaneously change the state of the system, changing the state of the system.

The modeling approach has several drawbacks when dealing with large systems.

To make the model look impressive, first of all, with every change, the system must take into account all the components of its general condition.

Secondly, with the above approach, information in systems disappears between random links.

Thirdly, the so-called asynchronous systems can cause undefined events at time intervals.

Petri nets and the above types of models are called asynchronous.

Causal relationships make it possible to more clearly describe the structural features of the system.

Asynchronous models of nonformal description of the case, in particular, Petri nets, must involve relationships of time (early, late, not at the same time, etc.), when it is convenient or accepted, but they represent a causal relationship. Great interplay of asynchronous systems, typically, has a complex dynamic structure.

The relationship between the two will be described more clearly if not immediate contacts are marked, or cases and situations in which the case can be realized. In this case, the conditions of implementation of the system of global situations are formed in the named local operations.

The term has its capacity. The term is not fulfilled (capacity is equal to 0), the term is fulfilled (capacity is equal to 1), and the term is fulfilled in n times (capacity is equal to n, where n- is a positive integer).

Most systems are suitable as discrete structures that consist of two elements: type of events and terms. Cases and terms in Petri nets, sets that do not intersect with each other, respectively, are called positions and transitions. Transitions are vertical lines and places with circles in a graphical representation of Petri nets [1, 2].

2.1 The relationship of petri nets, reachable states, and reachable trees

Definition 1: Petri nets are $M(C,\mu)$, where $C=(P,T,I,O)$ is the network structure and μ is the network condition. P is positions and T is transitions, which are finite sets. $I:T\to P^{\infty}, O:T\to P^{\infty}$ are input and output functions, respectively, where P^{∞} are all possible multisets (repetitive elements) of P. $\mu:P\to N_0$ is the function of condition, where $N_0=\{0,1,\cdots\}$ is the set of integers and included 0.

Now, we will define a function that determines the number of elements in their entering numbers in the collection [8]. X element enters into collection of B, which we will appoint as $\#(X, B)$ (called: X number in B). If we limit the number of elements in the collection so that $0 \leq \#(X, B) \leq 1$, then we will reach the idea of the set. Since $\#(X, B)$ function determines the X element entering collection of B, it follows that $\#(X, B) \geq 0$, the grouping of all the X and B. X element of the B collection, if $\#(X, B) > 0$, i.e. $X \in B_,$. Identically, if $\#(X, B) = 0$, then $X \notin B$.

Let us set empty collection of $\varnothing$, which has members (i.e., all $X : \#(X, B) = 0$). $|B|$ is the capacity of the entire number of elements entering B collection:

$$|B| = \sum_X \#(X, B).$$

Saying net state, we will understand the following:

$$(\mu(P_1), \mu(P_2), \dots, \mu(P_n)), \quad n = |P|, P = \{P_1, \dots, P_n\}$$

Suppose we have $M = (C, \mu)$.

We will say that in μ state $t_j \in T$ transition is allowed to implement if for $\forall P_i \in I(t_j)$ there is

$$\mu(P_i) \geq \#(P_i, I(t_j)).$$

Suppose in μ state t_j transition is allowed to implement and it is actually acted. In this case the net will appear in its new state, μ', which is solved in the following way:

$$\forall P_i \in P, \mu'(P_i) = \mu(P_i) - \#(P_i, I(t_j)) + \#(P_i, O(t_j))$$

Let us name $R(C, \mu_0)$ as the reachable state set:

1. $\mu_0 \in R(C, \mu_0)$,

2. If $\mu' \in R(C, \mu_0)$ and $\exists t_j \in T$ have transition in the way that $\delta(\mu', t_j) = \mu''$, then $\mu'' \in R(C, \mu_0)$.

3. Other states do not belong to $R(C, \mu_0)$. The $R(C, \mu_0)$ can be infinite μ'' marking covers μ' marking if

$$\mu'' \geq \mu'.$$

First, build a reachability tree. Then you need to look for the peak as follows. If there is no such peak, then the marking is not covered by any achievable marking, if it is located inside and gives an accessible marking that covers [14–16].

We construct the reachability tree of Petri nets in **Figure 1**. The state of this network is (1101), which shows the presence of tokens in the network at this moment. Tokens shown in **Figure 1**, which are depicted with small dots, correspond to the availability of resources. The network state changes due to the movement of tokens.

Let the states correspond to the vertices and transitions to the sides. The root corresponds to the first state of the network.

Figure 1 corresponds to **Figure 2**, in which the reachable tree is infinite. To make the tree finite, we impose restrictions. If any peak is blocked, we will call it a terminal. If there is a state in any peak and there is another peak in the tree with the

DOI: http://dx.doi.org/10.5772/intechopen.90275

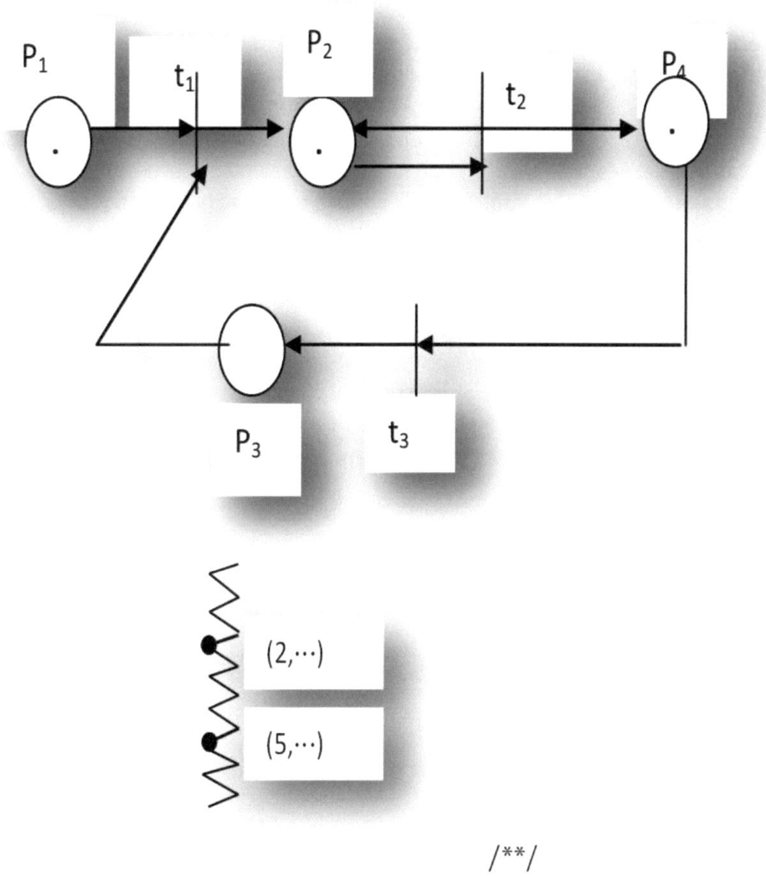

Figure 1.
An example of petri nets. The way in the tree.

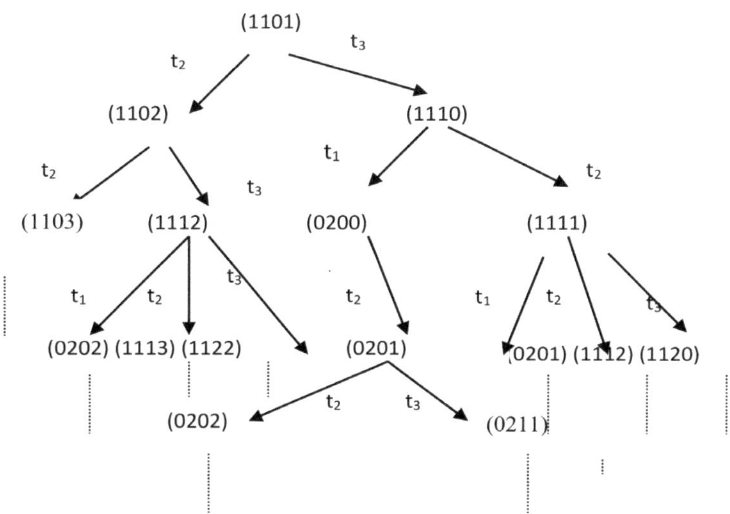

Figure 2.
The petri net reachable infinite tree.

same state that has already been developed, then we will call the new peak repeated and will not develop it.

If there is a path like / ** / in the tree, then the path through the second peak may repeat, and the states grow. Let us introduce the idea of infinite much as ω:

$\omega \geq \omega$, $\omega + a = \omega$, and $\omega - a = \omega$, where $a = \text{const}$: for example, instead of $(5, \cdots)$, we will write $(\omega, \cdots)$. In this case the tree will become as finite, and we will have loss of information [2].

Let us give several definitions, which will be used in the entire work.

Definition 2. A peak is called a boundary if it is in a processing state.

Definition 3. The peak is called a terminal if it does not contain a subtree.

Definition 4. The peak is called internal if it has already been processed.

Definition 5. The boundary peak is repeated if there is an internal peak with the same state.

A description of the structure of the reachability tree algorithm can be seen in an earlier published article [8].

With the help of this algorithm, we will build (as in **Figure 1**) the Petri net reachable tree (see **Figure 3**).

2.2 The algorithm for finding the minimum number of transitions in a coverage state

Consider the Petri net in **Figure 1** and the corresponding reachable tree TT (see **Figure 3**).

We note the set of states in Petri nets with P. Let T^* denote the succession of transitions from the root TT to y, the transition sequence with G the succession of the peaks in T^*.

Consider $\mu[x] = (0, 1, 15, 13)$ state. Let us find the y peak of this reachable tree for which the following inequality holds: $\mu[y] \geq \mu[x]$.

Assume that such peaks are $y_1, \ldots, y_m$. Let us choose one peak among the peaks on which we will use the algorithm.

For every y_i peak, we profile $\mu[y_i]$.

Suppose in $\mu[y_i]$ there is ω in $\mu[y_i]_{i1}, \ldots, \mu[y_i]_{ik}$. For each $\mu[y_i]$ we count $S = \sum_{j=i_1}^{i_k} z_j(t)$, where.

$z_j(t) = \#(P_j, I(t_k)), \forall t_k \in T^*$:

We take the y_i for which the S is the minimum. If for any peak, these numbers are equal, then we take the y_i in which T^* height is the minimum.

For example, $\mu[x]$, we will cover the following peak:

- $y_1 \ \mu[y_1] = (1, 1, \omega, \omega), \ T^* = \{t_2, t_3\}$.
- $y_2 \ \mu[y_2] = (0, 2, \omega, \omega), \ T^* = \{t_2, t_3, t_1\}$.
- $y_3 \ \mu[y_3] = (1, 1, \omega, \omega), \ T^* = \{t_2, t_3, t_2\}$.
- $y_4 \ \mu[y_4] = (1, 1, \omega, \omega), \ T^* = \{t_2, t_3, t_3\}$.
- $y_5 \ \mu[y_5] = (1, 1, \omega, \omega), \ T^* = \{t_3, t_2, t_3\}$.
- $y_6 \ \mu[y_6] = (0, 2, \omega, \omega), \ T^* = \{t_3, t_1, t_2, t_3\}$.

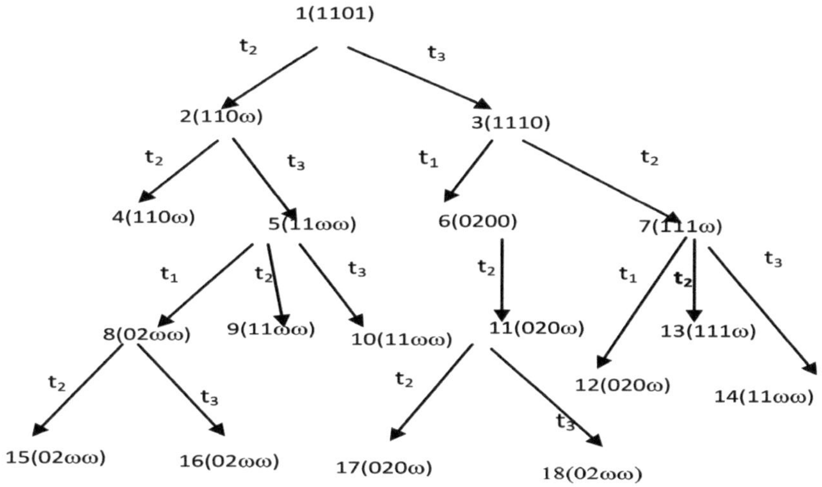

Figure 3.
The petri net reachable tree.

- $S(y_1) = 1 \quad S(y_4) = 2$
- $S(y_2) = 2 \quad S(y_5) = 2$
- $S(y_3) = 1 \quad S(y_6) = 3$

We found out that in minimum number, $S(y_1) = S(y_3)$, $|T^*(y_1)| = 2$, and $|T^*(y_3)| = 3 \Rightarrow$ need to take a y_1 peak. Choosing the appropriate peak coverage, we use the algorithm. Let it be a covering peak.

1. We choose the path connecting the root of the tree with$y$ and T^*; for our example t_2, t_3 let us mark $t'_i = t_j, 1 \le i \le |T^*|$, and $t_j \in T^*$. We get $t'_1 = t_2$, and $t'_2 = t_3$.

2. For each chosen transitions, t'_i is corresponded with a_i numbers in the following way:

   - If for t'_i transition $\exists 1 \le j \le |P|$ in the way that$\delta(\mu[y'], t'_i)_j = \omega, y' \in G$ then $a_i = \mu[x]_j$ in which case t'_i transition corresponds with P_j position.

   - If for the same t'_i transition $\exists 1 \le k \ne j \le |P|$ in the way that $\delta(\mu[y'], t'_i)_k = \omega$, then $a_i = \max\left\{\mu[x]_j, \mu[x]_k\right\}$.

Moreover, for t'_i transition we will correspond P_j and P_k positions. If instead of $t'_i\ \sigma = t'_{i_1} \dots t'_{i_k}\left(t'_{i_k} = t'_i\right)$ for $\exists 1 \le j \le |P|$ in the way that $\delta(\mu[y'], \sigma)_j = \omega, y' \in G$, then we will correspond a_i with σ and $a_i = \mu(x)_j$.

In this case, we will correspond σ with P_j position. In the opposite case, if there is no t'_i transition for $1 \le j \le |P|$ in the way that $\delta(\mu[y'], t'_i)_j = \omega$, then $a_i = 1$, in which case there is no related position for t'_i.

For example:

- $a_1 = 13 \, a_2 = 15$
- $t'_1 \sim P_4 \, t'_2 \sim P_3$.

Now, we will define the following action for a_i:

$$a_i = \begin{cases} \dfrac{a_i - n}{m}, & \text{if } (a_i - n)\text{modm} = 0 \\ \left\lceil \dfrac{a_i - n}{m} \right\rceil + 1, & \text{if } (a_i - n)\text{modm} \neq 0 \end{cases}$$

where n to t'_i or in σ corresponding P_j position, the number of tokens are in their first position, and m from t'_i or σ to the number of the arrows in the state: $\#(P_j, O(t'_i))$.

If for t'_i transition $P_1, \cdots, P_k$ positions correspond, then we will take the P_1 position for which $\mu[x]_1 = a_i$. In this case $n = \mu_0[P_1]$, and $m = \#(P_j, O(t'_i))$.

If there is no corresponding position for t'_i transition, then we will leave a_i to remain the same.

For example:

$$a_1 = (a_1 - 1)/1 = (13 - 1)/1 = 12$$
$$a_2 = (a_2 - 0)/1 = (15 - 0)/1 = 15.$$

Let us mark $b_i^1 = a_i$. For example:

$$b_1^1 = a_1 = 12$$
$$b_2^1 = a_2 = 15.$$

2.2.1 Cumulative move

We will take T^* last transition or the succession of transition, fix it and mark as t_α.

t_α corresponding b_i^1 is marked as α which we also fix. The fixed b_i^j does not change in the next moves.

We consider all T^* items from the right to left, starting from t_α.

Suppose t'_k is the considered transition or the transition succession and P_1 is the corresponding position of t'_k.

If $P_1 \in I(t_\alpha)$, then t'_k corresponding b_i^j in the next move will get the following value: $b_i^{j+1} = b_i^j + \alpha \cdot l$, where $_l$ from P_1 position t_α is the number of arrows.

Suppose t'_k corresponds with $P_1, \cdots, P_l$ positions. If $\exists 1 \leq j \leq l$ in the way that$P_j \in I(t_\alpha)$, then $b_i^{j+1} = b_i^j + \alpha \cdot l$, where $l = \#(P_j, I(t_\alpha))$. In the opposite case, $b_i^{j+1} = b_i^j$.

After that, we fix t_α the previous transition action and denote it as t_α.

We denote the new t_α corresponding to b_i^j as α and go to the second step again.

It follows that for each transition or sequence of transitions, there will be a correspondingly fixed number b_i^j, which will mark the number of implementation of the transition or sequence of transitions. For example:

t'_1	t'_2
12	15
27	15

The above shows that the t'_1 transition must be implemented for 27 times and t'_2 transition for 15 times.

Given our denote, we get that state $\mu[x] = (0, 1, 15, 13)$ covers state $\mu[y] = (1, 1, \omega, \omega)$. To achieve the goal, we need to implement the t_2 transition 27 times and the t_3 transition for 15 times.

Let us assign $t_s(y) = \sum_{i=1}^{l} b_i^j, l = |T^*(y)|$. Which means $t_s(y)$ is y the number of enabled transitions.

Lemma 1. Suppose there are y_1 and y_2 peaks in the way that $\mu(y_1) \geq \mu[x] \& \mu(y_2) \geq \mu[x]$. In that case $t_s(y_1) \leq t_s(y_2)$.

Proof: We have.

$$S(y) = \sum_{j=i_1}^{i_k} z_j(t), \text{ where } z_j(t) = \#(P_j, I(t_k)), \forall t_k \in T^*.$$

$$t_s(y) = \sum_{i=1}^{k} b'_i, \text{ where } k = |T^*(y_1)|.$$

In this number some b'_is are equal to 1. Without breaking the sense we will suppose that the first d' number of b'_is is equal to 1. We will get

$$t_s(y_1) = d' + \sum_{i=d'+1}^{k} b'_i = d' + \sum_{i=d'+1}^{k} (\mu[x]_l + m' \cdot b')$$

We have $t_s(y_2) = \sum_{i=1}^{k'} b''_i$, where $k' = |T^*(y_2)|$.

Suppose for b''_is, number of d'' is equal to 1. Moreover $d'' \leq d'$, as $S(y_1) < S(y_2)$,

$$t_s(y_2) = d'' + \sum_{i=d''+1}^{k'} (\mu[x]_l + n''_i \cdot b'') \geq d' + \sum_{i=d'+1}^{k} (\mu[x]_l + n'_i \cdot b') = t_s(y_1)$$
$$\Rightarrow t_s(y_1) \leq t_s(y_2)$$

The lemma is proven.

Lemma 2. Suppose y_1 and y_2 are covering peaks. There is $S(y_1) = S(y_2) \& |T_1^*| < |T_2^*|$: in this case $t_s(y_1) < t_s(y_2)$.

Proof:

$$t_s(y_1) = d' + \sum_{i=d'+1}^{k} b'_i = d + \sum_{i=d'+1}^{k} (\mu[x]_l + n'_i \cdot b') \underset{d' < d''}{<} d'' + \sum_{i=d''+1}^{k'} (\mu[x]_l + n''_i \cdot b'')$$
$$= t_s(y_2).$$

The lemma is proven.

Theorem. Through the abovementioned number of covering state, transition algorithm is in its minimal state.

Proof: Let y be the covering peak in our algorithm and $t'_1, \cdots, t'_k$ the succession of transitions. It must be shown that the number of $t'_1, \cdots, t'_k$ move is in minimal state. For this we need to show that $\nexists y'$ covering peak has less number of transitions than the number of $t'_1, \cdots, t'_k$.

Let us consider two cases:

1.$y \neq y'$

Suppose the transition number of y' is less than $t'_1, \cdots, t'_k$ implementation number. According to the algorithm: $S(y) < S(y')$ or

$$S(y) = S(y')\&|T_1^*| < |T_2^*|.$$

If $S(y) < S(y') \Rightarrow$ according to Lemma 1: $t_s(y) \leq t_s(y')$. We've come into a controversy.

If $S(y) = S(y')\&|T_1^*| < |T_2^*| \Rightarrow$ according to Lemma 2: $t_s(y) \leq t_s(y')$. We've come into a controversy.

2.$y = y'$

Suppose succession transitions of y' is $s_1, \cdots, s_r$. As the tree does not contain any cycle, $y = y' \Rightarrow t'_1, \cdots, t'_k$ and $s_1, \cdots, s_r$ are the same $\Rightarrow t_s(y) \leq t_s(y')$. The theorem is proven.

2.3 Conclusion

The proven theorem and research reveal some important features of Petri nets from the point of view of optimization, that is, if the idea of Petri nets is used in technical devices, then the idea of sequential transitions save resources and time.

3. Interrelation of languages of colored Petri nets and some traditional languages

Definition. The mathematical definition of colored Petri net: CPN is a nine-tuple $CPN = (\Sigma, P, T, A, N, C, G, E, I)$, where:

$\sum$ is a finite set of non-empty types called color sets [17].
P is a finite set of places which are depicted as ovals/circles.
T is a finite set of transitions which are depicted as rectangles.
A is a finite set of arches which are depicted as directed edges; moreover.

$$P \cap T = P \cap A = T \cap A = \varnothing.$$

N is a node function, $A \to P \times T \cup T \times P$.
C is a color function, $C : P \to \Sigma$.
G is a guard function. It is defined from T into expressions such that

$$t \in T : [Type(G(t)) = B\&Type(Var(G(t))) \subseteq \Sigma].$$

E is an arc expression function, which is defined as follows:

$$\forall a \in A : [Type(E(a)) = C(p)_{MS}\&Type(Var(E(a))) \subseteq \Sigma],$$

I is an initialization function [3–6, 9, 10],

$$\forall p \in P : [Type(I(p)) = C(p)_{MS}].$$

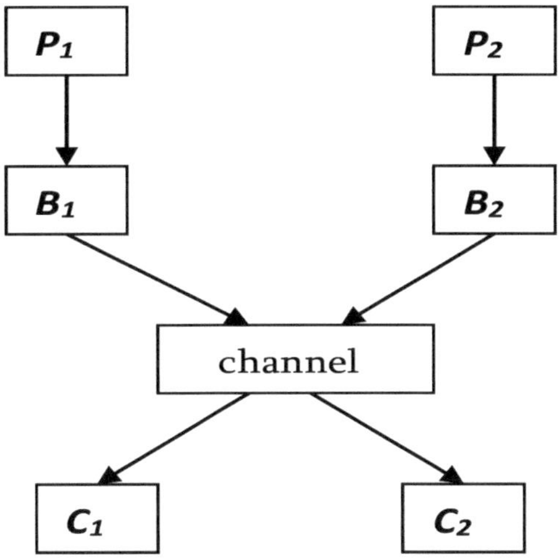

Figure 4.
The consumers' process with the common usage and buffer is an action.

The distribution of tokens, called marking, determines the state of the simulated system. The dynamic behavior of CPN is due to the triggering of a transition that transfers the system from one state to another. A transition is enabled if the associated arc expressions of all input arches can be evaluated as a multi-set, which is compatible with the current tokens in corresponding input places, and its guard is satisfied. After the transition is triggered, tokens are removed from the input places, respectively, by the specified expression of the arches of all incoming arches, and tokens are placed in the output places, respectively, by the specified expressions of the outgoing arches [3–6, 17].

3.1 The example of modeling consumers' process with CPN

Let us suppose that there are two processes of producers and consumers [1, 9].

The following picture shows the process diagram (**Figure 4**).

There is a distribution problem in the described system. To use the channel, the pair(P_1, C_1) must have priority toward (P_2, C_2) in the sense of using the channel. This is described as follows: while the buffer is not empty, the channel cannot report data from the buffer to the consumer. It is impossible to solve this problem with the help of classical Petri nets, since it is permissible in nature. The proof of this fact is described in the literature [1].

To solve that problem, it is needed to extend Petri net's several properties in such a manner that the proposed properties are headed toward the opportunity of checking the zero in Petri nets [13].

3.2 Declaration

```
Color E = {e};
Color Control = {0;1};
Color S = product E*Control;
Var ct:Control;
```

The CPN (**Figure 5**) is the model of the solved problem of priority usage [17, 19].

Declaration:
Color E={e};
Color
Control={0;1};
Color S=product
E*Control;
Var ct:Control;

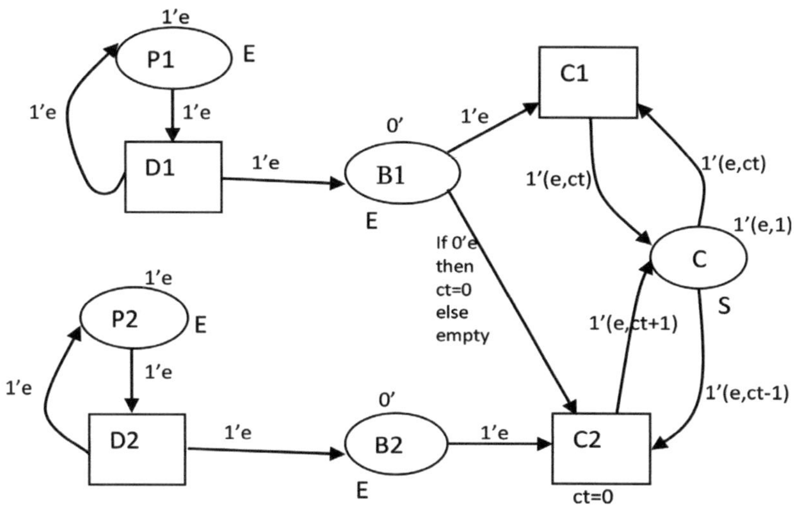

Figure 5.
The modeling of consumer problem with colored petri net.

3.3 The modeling of context-free languages with colored Petri nets: the diagram of interrelation of colored Petri nets and traditional languages

It is known that the class of regular languages is one of the most studied simple classes of formal languages and any regular language is the language of Petri nets [2, 18].

There are context-free languages that are not languages of Petri nets. Such examples of the context-free languages are $\{\omega\omega^R/\omega \in \Sigma^*\}, L^* = L \cup LL \cup LLL...$ (in particular, $\{a^n b^n/n > 1\}$).

The noted fact shows the limitation of Petri nets as a mechanism that generates languages [2].

In Petri nets one can only remember a sequence of limited length (similar to finite automata) [2].

It is clear that Petri nets do not possess the "capacity of pushdown memory" necessary for generating context-free languages. The relationship of the languages of Petri nets with other classes of languages (Venn diagram) is shown in **Figure 6** [2, 10].

3.4 Results

A model of the $L^* = L \cup LL \cup LLL...$ language (Klins' star) is constructed using colored Petri nets, in particular $\{L = a^n b^n/n \geq 1\}$.

Colored Petri net (**Figure 7**) generates such a language, which proves that the colored Petri net is a more powerful tool than the classical Petri nets. The following declaration is for the concept of data types.

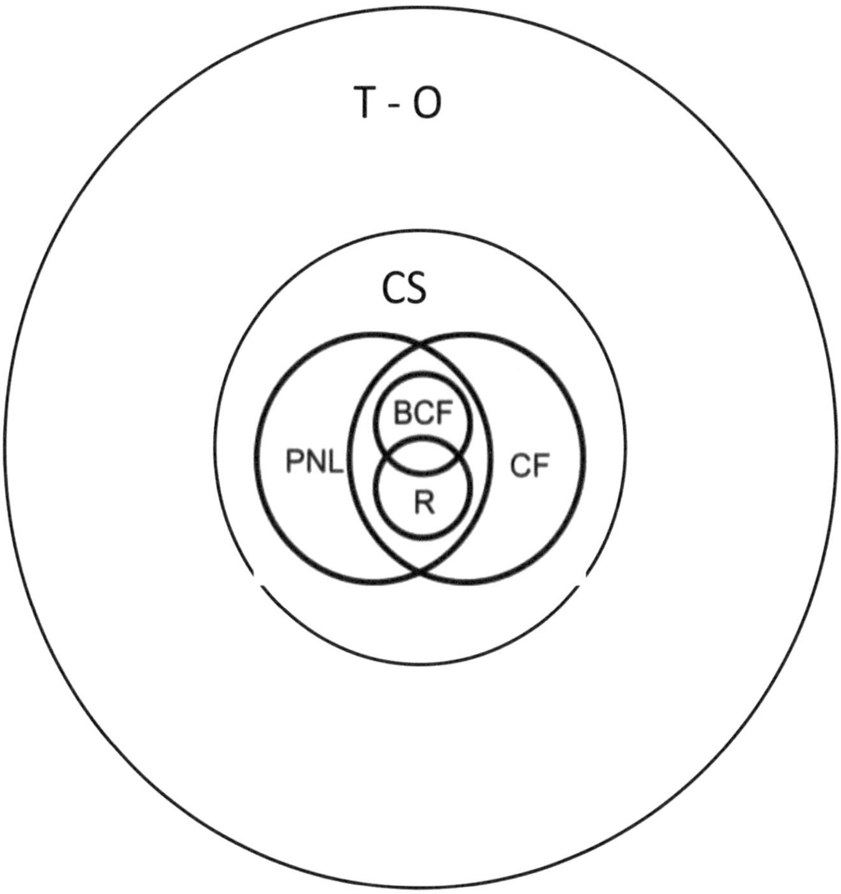

Figure 6.
Interrelation of petri nets and traditional languages (T-0, the general type of languages; CS, context-sensitive languages; PNL; petri net languages; CF, context-free languages; BCF, bonded context-free languages; R, regular languages).

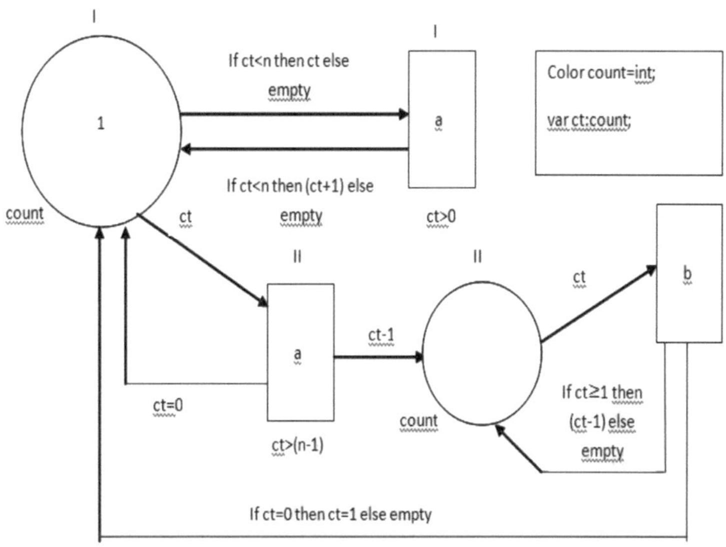

Figure 7.
Modeling $L^* = L \cup LL \cup LLL \dots$ *language by colored petri net.*

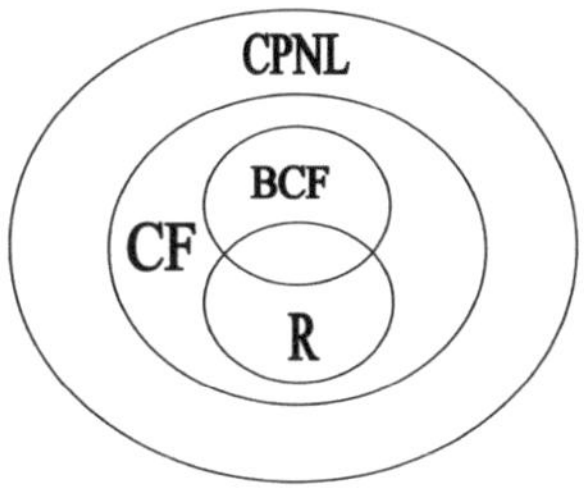

Figure 8.
Interrelation of colored petri nets and traditional languages. (CPNL, language of colored petri net).

The operation of the colored Petri net shown in **Figure 7** is described in more detail in the literature [3, 10].

The colored Petri net (**Figure 7**), which is built for $L^* = L \cup LL \cup LLL \ldots$ language, suggests the following relationship between the languages of the colored Petri nets with some classes of traditional languages (see **Figure 8**) [10].

The Venn diagram, modified by the author (**Figure 8**), shows the relationship between the languages of the colored Petri nets and some traditional languages. This fact illustrates that the language class of colored Petri nets includes an entire class of languages without context.

4. On a solution to the cigarette smoker's problem with colored Petri nets

In 1971 Patil proved that P and V actions have insufficient capacity for resolving synchronization issues. His proposed solution to model problem is called smoking a cigarette [9].

The actions of the smokers without the coordination are as follows.

Let X be the smoker with tobacco, Y the smoker with paper, Z the smoker with matches, and A the agent (see **Table 1**).

It is proven that the problem of smokers has no solution using semaphores [9].

Patil showed that there is no sequence of P and V actions to correctly solve the problem [1, 2]. Modeling the problem using the classical Petri net, we get an inactive network. Since all tokens in classical Petri nets are of the same type, the ingredients will not differ from each other.

The author simulated a problem with the colored Petri net (see **Figure 9**) [3–6, 9, 18, 19].

The operation of the colored Petri net shown in **Figure 9** is described in more detail in the literature [9].

If we were to represent this problem using the classical Petri net, then we need to use three transitions instead of one **T** transition. It also means that minimization of the network is ensured, which implies a reduction in costs due to the reduction of arches in positions and transitions.

Processes A_X	Processes A_Y	Processes A_Z
Pick up the paper	Pick up the tobacco	Pick up the tobacco
Pick up the match	Pick up the match	Pick up the paper
Roll the cigarette	Roll the cigarette	Roll the cigarette
Light the cigarette	Light the cigarette	Light the cigarette
Smoke the cigarette	Smoke the cigarette	Smoke the cigarette
Return to A_X	Return to A_Y	Return to A_Z

Table 1.
The actions of the smokers.

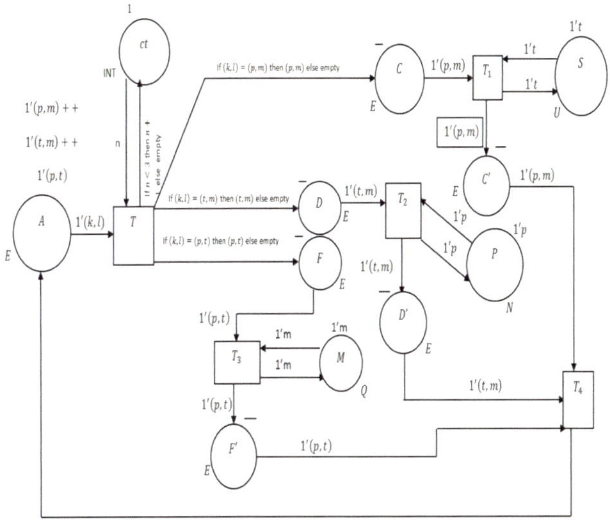

Figure 9.
The modeling of cigarette smoker's problem with colored petri nets.

4.1 Declaration

Color INT = integer;
Color U = t;
Color N = P;
Color Q = m;
Color E = {Product N*Q OR Product U*Q OR Product N*U};
Var(K, l) : E;
n : INT;

4.2 Conclusion

In the problem, we identify certain advantages of colored Petri net to P and V operations and classical Petri net with the synchronization problem. The mentioned studies allow identification of synchronization modeling opportunities with the help of colored Petri net.

Author details

Goharik Petrosyan
Plekhanov Russian University of Economics Yerevan Branch, Member of Armenian Mathematical Union, ASPU, ISEC of NAS RA, Yerevan, Armenia

*Address all correspondence to: petrosyan_gohar@list.ru; petrosyangoharik72@gmail.com

IntechOpen

© 2019 The Author(s). Licensee IntechOpen. This chapter is distributed under the terms of the Creative Commons Attribution License (http://creativecommons.org/licenses/by/3.0), which permits unrestricted use, distribution, and reproduction in any medium, provided the original work is properly cited.

References

[1] Tadao M. Petri nets: Properties, analysis and applications. Proceedings of the IEEE. 1989;77(4)

[2] Peterson J. Petri Net Theory and the Modelling of Systems. Prentice Hall; 1981. ISBN 0-13-661983-5

[3] Jensen K, Rozenberg G, editors. High-Level Petri Nets. Berlin: Theory and Application, Springer-Verlag; 1991. pp. 44-122

[4] Jensen K. Colored Petri Nets: Basic Concepts, Analysis Methods and Practical Use. Berlin: Springer-Verlag; 1992

[5] Jensen K. Colored Petri Nets: Basic Concepts, Analysis Methods and Practical Use. Vol. 1–3. Springer; 1996

[6] Jensen K. Colored petri nets: Basic concepts, analysis methods and practical use. In: Basic concepts. Monographs in Theoretical Computer Science. Vol. 1–3. Berlin, Germany: Springer–Verlag; 1997

[7] Raising W, Rosenberg G, editors. Lecture notes on petri nets. Parts I and II. In: Lecture Notes in Computer Sciences. Vol. 1491–1492. Springer-Verlag; 1998

[8] Petrosyan GR. Description of the algorithm for finding the shortest sequence of transitions in a Petri net. Multidisciplinary Scientific Edition International Academy Journal Web of Scholar. 2017;**5**(14):20-25. Available at: http://webofscholar.com/ ISSN 2518-167X Founder – RS Global Media LLC, Kiev, Ukraine

[9] Petrosyan GR. The modelling of the synchronization problem with Colored petri nets. International Journal of Electronic Engineering and Computer Science. 2016;**1**(2):56-60 Available at: http://www.aiscience.org/journal/ijeecs

[10] Petrosyan GR, Avetisyan AM, Ter-Vardanyan LA. Interrelation of languages of colored petri nets and some traditional languages. Open Journal of Modelling and Simulation. 2013;**1**:27-29. DOI: 10.4236/ojmsi.2013.13005. Published Online July 2013 (http://www.scirp.org/journal/ojmsi)

[11] Westergaard M, Kristiansen L. The access/CPN framework: A tool for interacting with the CPN tools simulator. In: Proc. of 30th International Conference on Applications and Theory of Petri Nets (Petri Nets 2009). Lecture Notes in Computer Science 5606. Berlin: Springer-Verlag; 2009. pp. 313-322

[12] Jensen K, Kristiansen L, Wells L. Colored petri nets and CPN tools for modelling and validation of concurrent systems. International Journal on Software Tools for Technology Transfer (STTT). 2007;**9**(3–4):213-254

[13] Petrosyan GR, Ter-Vardanyan LA, Gaboutchian AV. Modeling of biometric identification system using the colored petri nets. The International Archives of the Photogrammetry, Remote Sensing and Spatial Information Sciences. 2015; **XL-5/W6** Photogrammetric techniques for video surveillance, biometrics and biomedicine, 25–27 May 2015, Moscow, Russia

[14] Knut D. The Art of Programming. Vol. 1–3. Moscow: Mir; 1976

[15] Orlov S. Technology of Software Development, Textbook for Universities. Petersburg; 2002

[16] Gordeev A, Molchanov A. System Software, Textbook. St. Petersburg; 2002

[17] Jensen K, Kristiansen L. Coloured Petri Nets—Modeling and Validation of Concurrent Systems. Berlin: Springer-Verlag; 2009

[18] Alfred A, Jeffrey U. Theory of Parsing, Translation, & Compiling. Vol. 1-2. Prentice Hall; 1973

[19] Ullman J. Elements of ML Programming. Upper Saddle River: Prentice-Hall; 1998

Chapter 8

Line Impedance Emulator: Modeling, Control Design, Simulation and Experimental Validation

Marwa Ben Saïd-Romdhane, Sondes Skander-Mustapha and Ilhem Slama-Belkhodja

Abstract

The variation of line impedance has always been a great concern for grid operators and industrial users. The problem is that the reliability and quality of the supplied power are influenced by this variation. Indeed, several standards and grid requirements fix strict rules and rigorous standards when connecting or disconnecting from the public grid. In this context, this chapter proposes a full study of a line impedance emulator, which includes the power design and the control. The line impedance emulator is useful for small scale laboratories that develop distributed energy generation. Developed line impedance emulator is based on a three-phase power converter. For these converters, different controls are applied, including proportional integral and resonant controllers. For the generation of voltage reference values that correspond to expected line impedance, two algorithms are studied, namely, trigonometric functions-based algorithm and voltage drop-based algorithm. The theoretical study is supported by simulation and experimental results.

Keywords: power quality, distributed energy generation, microgrid, line impedance emulation, resonant controller

1. Introduction

Nowadays, with the tremendous increase of distributed energy generation (DEG), the concept of power quality (PQ) has become a growing concern for grid operators around the world [1–4]. Many research teams working on this topic are developing small or large-scale DEG laboratories (**Figure 1**) [3–9] as well as algorithms for critical situations is the grid emulator. This grid emulator is also used to confirm the compliance with standards and different grid codes [10–13].

This chapter covers one of the functionalities of the grid emulator, which is the line impedance emulation. Indeed, line impedance deviation can be caused by several circumstances, such as, a remote grid fault, or a connection disconnection of a large load in the distribution network [14].

IntechOpen

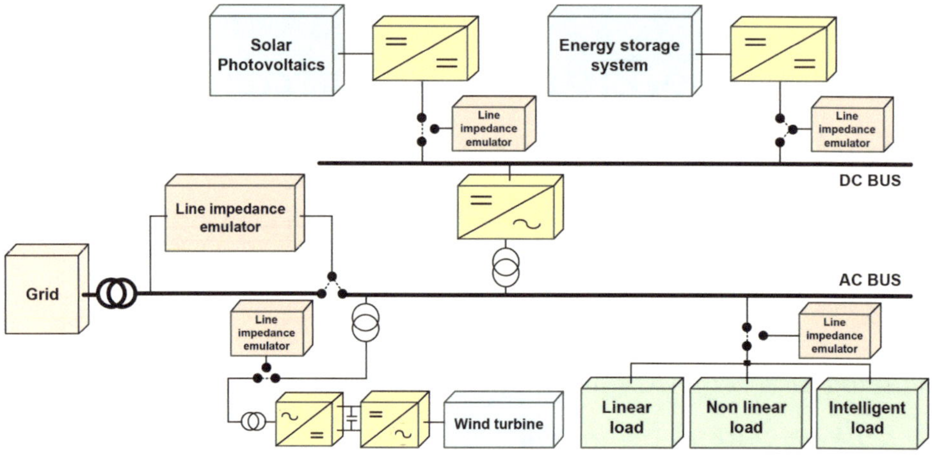

Figure 1.
Example of a microgrid including line impedance emulators.

The line impedance variation is able to considerably affect reactive power sharing between parallel loads [15, 16], and it can also induce operation instability in case of standalone microgrid [17, 19]. In addition, line impedance value has an influence on the quality of voltage and line current in the point of common coupling of the microgrid [19]. In another hand, tests introducing line impedance variation are used for the compliance with many relevant standards especially those dealing with anti-islanding.

This chapter explains in details the steps of the line impedance emulator design based on power converters. Regarding line impedance emulation algorithm, reference voltage values are deduced in view of the phase shift with the input AC grid voltage, according to the equipment under test (EUT) active and reactive power. Presented emulator guarantee flexible tests with decoupled variation range of impedance component.

This chapter first outlines modeling of line impedance emulator, followed by a description of the control methodology for the overall, simulation results and experimental validation are then developed.

2. Line impedance emulator presentation

The line impedance emulator is installed between the grid and the EUT and used for the emulation of variable line impedance. The structure of the studied line impedance emulator system is shown in **Figure 2**. It incorporates two power converters joined by dc-link capacitor: an EUT side converter (EsC) and a grid side converter (GsC). The GsC and the EsC are AC/DC and DC/AC converters, respectively. To mitigate switching harmonics, an LCL filter is employed at the output of the EsC. The EsC control aims to maintain the voltage through the LCL filter capacitor $V_{c(abc)}$ equal to the programmed references. The GsC has the intention of regulating the system power factor (PF) and the voltage at the DC bus V_{dc}. As presented in **Figure 2**, the line impedance emulator output $V_{out(abc)}$ is equal to $V_{c(abc)}$, while its output $V_{in(abc)}$ is considered comparable to the grid voltage $V_{g(abc)}$.

The flowchart of the line impedance emulator process is given by **Figure 3**. The first step of this flowchart consists in initializing the different functions and the microcontroller peripherals such as the ADC, Timers and the General Purpose Input/Output (GPIO) as well as the analog-to-digital conversion of the measured

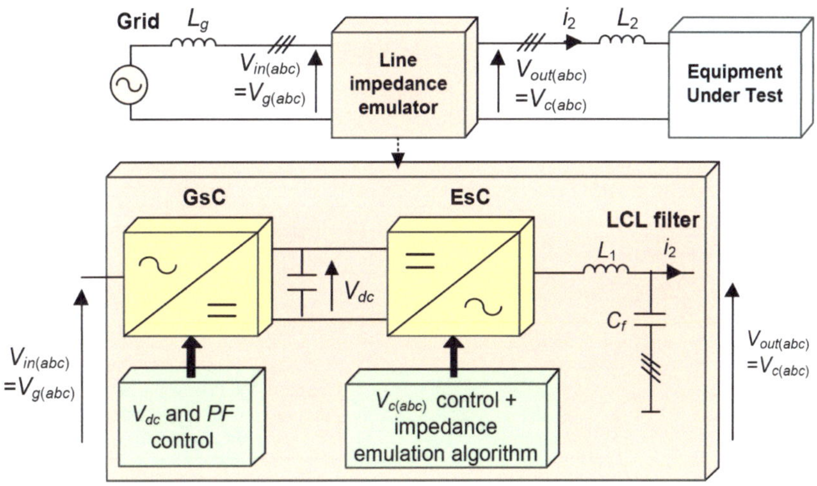

Figure 2.
Power converter-based three-phase line impedance emulator.

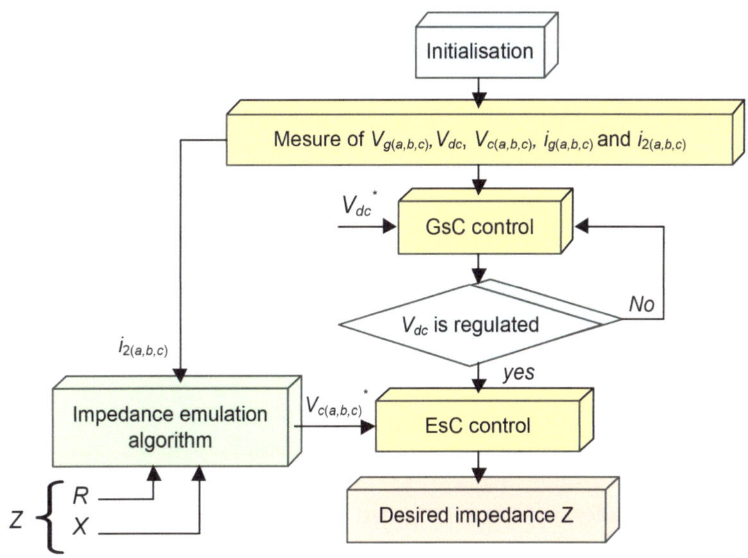

Figure 3.
Line impedance emulator process flowchart.

voltages and currents. The next step is to control of the GsC. The objective of this control is voltage at the DC bus regulation. In parallel with these steps, the impedance emulation algorithm provides the capacitor voltage references $V_{c(a,b,c)}^*$ according to programmed impedance. Once V_{dc} is equal to its reference and the capacitor voltage references $V_{c(a,b,c)}^*$ are generated, the operator proceeds to the control of the EsC. The desired line impedance is consequently achieved.

Figure 4 summarizes the different steps of the line impedance emulator design. As mentioned, the first step consists in modeling the two power converters of the line impedance emulator giving the system equations and transfer functions. After that, the operator selects the appropriate control converters control in terms of dynamic response, THD value, steady state error and sensitivity to perturbation and parametric variation. In this chapter, the control of the line impedance emulator converters employed resonant controllers and PI regulators. This choice is due to

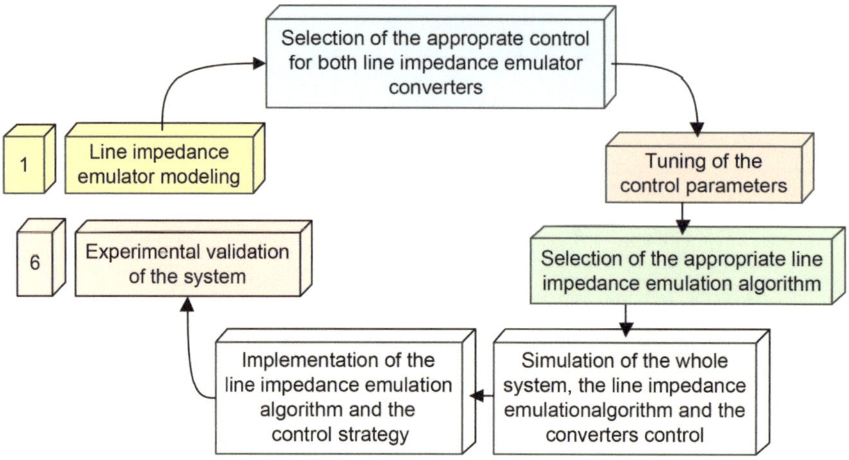

Figure 4.
Methodology of the design of a line impedance emulator.

their simple use (tuning parameters and implementation), while ensuring simultaneously acceptable dynamic response, THD value and steady state error. Then, based on the obtained system transfer functions, the control parameters are deduced. After that, the operator should select the appropriate line impedance emulator algorithm. In this chapter, two impedance emulator algorithms will be presented. The next step of the design methodology consists in simulating the whole system including the power converters, the control strategy and the line impedance emulation algorithm. When the simulation results verify the proper system operation, the control will be implemented on a digital board. The last step of the design methodology consists in the experimental validation of the line impedance emulator.

3. Line impedance emulator modeling

The GsC power circuit single phase representation is depicted on **Figure 5**, where L_g denotes the grid impedance. According to this figure, the GsC electric equation in the *abc* reference frame is given by Eq. (1).

$$U_i = V_g - L_g \frac{di_g}{dt} \tag{1}$$

The EsC power circuit single phase representation is given by **Figure 6**. Based on this Figure, the equations related to the EsC are given by Eq. (2), Eq. (3), Eq. (4) and Eq. (5). The obtained single phase simplified block diagram of the LCL-EsC is depicted on **Figure 7**.

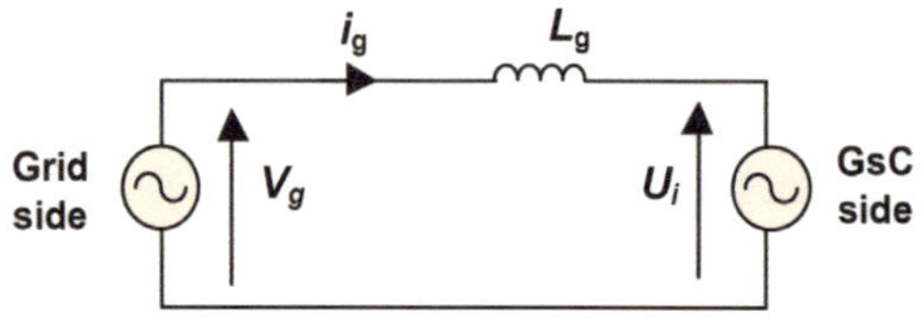

Figure 5.
GsC power circuit.

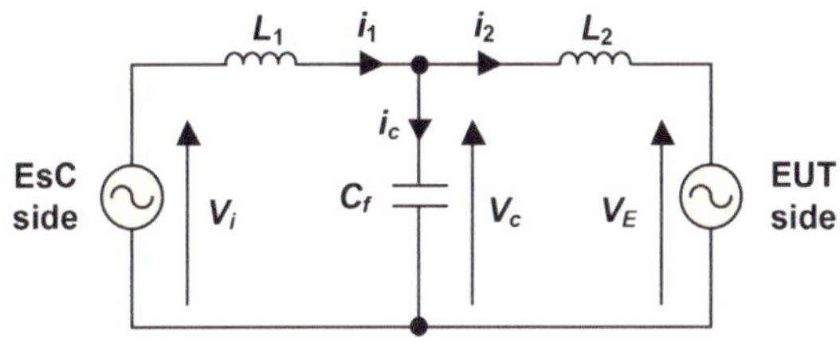

Figure 6.
EsC power circuit single phase representation.

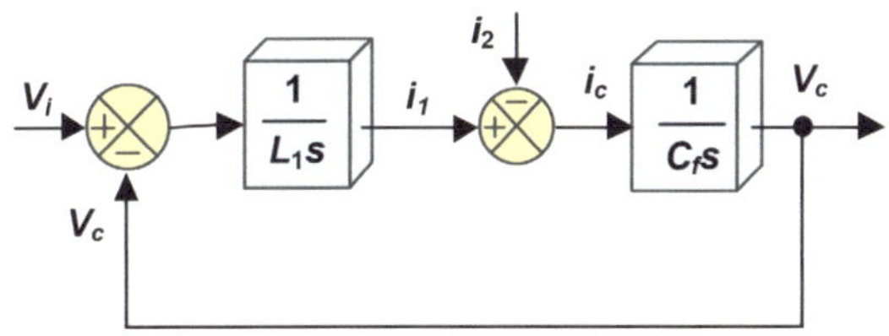

Figure 7.
LCL-EsC simplified block diagram.

$$i_1 = \frac{V_i - V_c}{sL_1} \tag{2}$$

$$i_1 = i_2 + i_c \tag{3}$$

$$V_c = \frac{i_c}{sC_f} \tag{4}$$

$$i_2 = \frac{V_c - V_E}{sL_2} \tag{5}$$

4. Line impedance emulation control

4.1 Grid side converter control

Figure 8 shows the GsC control. It incorporates two control loops. The internal loop controls in the *abc* reference frame the grid currents $i_{g(abc)}$ and it is based on

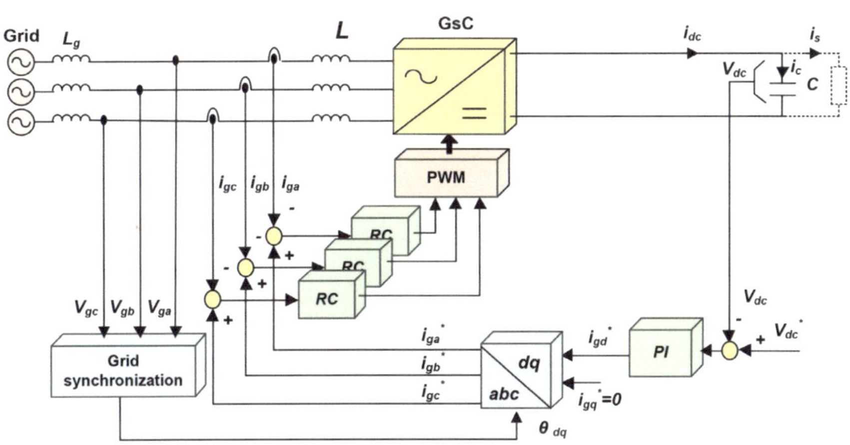

Figure 8.
Block diagram of GsC control.

resonant controller. The external loop regulates, via a PI regulator, the voltage at the DC bus V_{dc} and provides grid current reference on d axis i_{gd}^*. The grid current reference on the q axis i_{gq}^* is selected to have the desired PF. For the abc grid current reference components $i_{g(abc)}^*$, they are obtained via the application of Park transformation to i_{gd}^* and i_{gq}^*. In the following, the tuning of the PI and the resonant controller parameters will be detailed.

4.1.1 Tuning of the PI regulator of the voltage at the DC bus

Based on **Figure 9**, the current i_{dc} at the output of the GsC is expressed as in Eq. (6). By applying the Laplace transform to Eq. (6), Eq. (7) is obtained.

$$i_{dc} = i_c + i_s = C\frac{dV_{dc}}{dt} + i_s \tag{6}$$

$$V_{dc} = \frac{1}{Cs}(i_{dc} - i_s) \tag{7}$$

Since the current i_{dc} is instantaneously equal to $\pm i_g$ and the current regulation loop time constant is insignificant compared to the one of the DC bus voltage regulation loop, **Figure 9** gives simplified DC bus voltage regulation loop block diagram.

The transfer function of the PI regulator is given by Eq. (8). Based on this equation and neglecting the load current i_s, the closed-loop transfer function of the V_{dc} control is given by Eq. (9).

$$G_c(s) = \frac{i_{dc}^*}{\Delta V_{dc}} = K_{pdc} + \frac{K_{idc}}{s} \tag{8}$$

$$\frac{V_{dc}}{V_{dc}^*} = \frac{\frac{K_{pdc}}{C}s + \frac{K_{idc}}{C}}{s^2 + \frac{K_{pdc}}{C}s + \frac{K_{idc}}{C}} = \frac{\frac{K_{pdc}}{C}s + \frac{K_{idc}}{C}}{s2 + 2\xi_c\omega_{nc}s + \omega_{nc}^2} \tag{9}$$

The transfer function of Eq. (9) is a second-order system whose denominator can be written in the canonical form of a second-order system given by the right-hand side of Eq. (9). By identifying the terms of Eq. (9), the obtained transfer function is characterized by a damping ratio ξ_c and a natural frequency of oscillation ω_{nc} that satisfy Eq. (10) and Eq. (11).

$$2\xi_c\omega_{nc} = \frac{K_{pdc}}{C} \tag{10}$$

$$\omega_{nc}^2 = \frac{K_{idc}}{C} \tag{11}$$

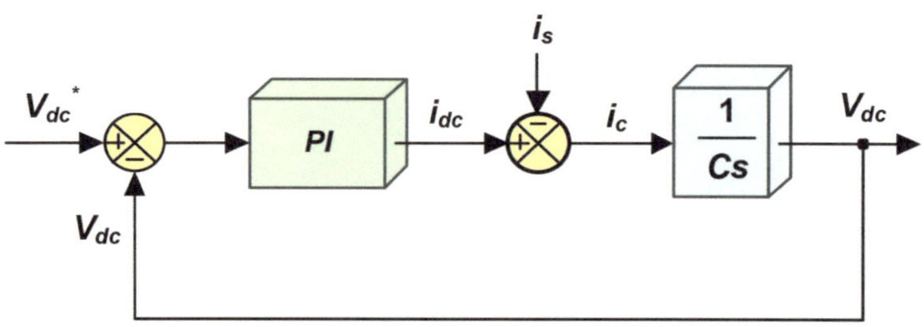

Figure 9.
DC bus voltage regulation loop simplified block diagram.

Then, the form and the dynamics of the response of the DC bus voltage V_{dc} are imposed by setting the natural frequency of the oscillations ω_{nc} and a damping coefficient ξ_c. Thus, the gains K_{pdc} and K_{idc} can be obtained based on equations Eq. (12) and Eq. (13).

$$K_{pdc} = 2C\xi_c\omega_{nc} \tag{12}$$

$$K_{pdc} = C\omega_{nc}^2 \tag{13}$$

4.1.2 Tuning of the resonant controller of the grid side current

The use of the PWM makes it possible to have a fundamental of the voltage U_i equal to its reference U_i^*. Thus, based on Eq. (1), we obtain the simplified single-phase block diagram the grid side regulation loop given by **Figure 10**.

Considering **Figure 10**, the closed-loop system transfer function (T_{cig}) is given by Eq. (14).

$$T_{cig}(s) = \frac{i_g(s)}{i_g^*(s) - i_g(s)} = \frac{K_{pig}s^2 + K_{iig}s + K_{pig}\omega_0^2}{L_g s^3 + K_{pig}s^2 + (L_g\omega_0^2 + K_{iig})s + K_{pig}\omega_0^2} \tag{14}$$

For the synthesis of the resonant controller parameters, we consider the pole placement method and more precisely the Naslin criterion [20–21]. The n order polynomial of this criterion is expressed by Eq. (15).

$$P_{Naslin}(s) = n_0\left(1 + s\tau + s^2\left(\frac{\tau^2}{\alpha}\right) + s^3\left(\frac{\tau^3}{\alpha^3}\right) + \dots\dots + s^n\left(\frac{\tau^n}{\alpha^{n(n-1)/2}}\right)\right) \tag{15}$$

From Eq. (14), we deduce the system characteristic polynomial given by Eq. (16).

$$P_{ig}(s) = L_g s^3 + K_{pig}s^2 + (K_{iig} + L_g\omega_0^2)s + K_{pig}\omega_0^2 \tag{16}$$

The identification between the system characteristic polynomial P_{ig} and the second order Naslin polynomial makes it possible the deduction of resonant controller parameters K_{pig}, et K_{iig} as shown in Eq. (17) and Eq. (18).

$$K_{pig} = L_g\frac{\alpha^2}{\tau} \tag{17}$$

$$K_{iig} = L_g\left(\frac{\alpha^3}{\tau^2} - \omega_0^2\right) = L_g(\alpha^2 - 1)\omega_0^2 \tag{18}$$

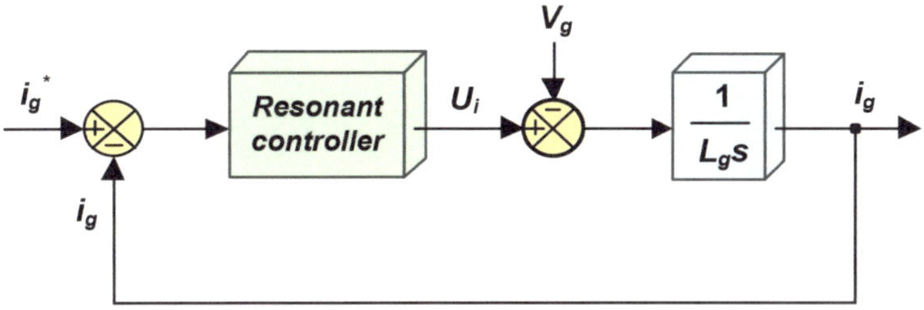

Figure 10.
Grid current regulation loop simplified block diagram.

4.2 EUT side converter control

The control based on resonant controller for the EsC is depicted on **Figure 11**. This control includes an external and an internal loops. The external one controls the voltages through the filter capacitor $V_{c(a,b,c)}$. The internal one controls the inverter side current $i_{1(a,b,c)}$ and generates then the inverter voltages references $V_{i(a,b,c)}$. For the external loop, a resonant controller is adopted. For the internal loop, the resonant controller is replaced by a constant gain (G) in order to ensure a faster loop than the external one. In the following, the tuning of the resonant controller parameters will be detailed and discussed in order to ensure good control performances.

4.2.1 Tuning of the resonant controller of the voltage through the LCL filter capacitor

For reasons of simplification, it is assumed that the internal loop of the current is faster than the external loop of the voltage. Thus, we can approximate it equal to the unity by associating the PWM function. Consequently, the block diagram of the voltage regulation loop is given by **Figure 12**.

Hence, the closed loop system transfer function (T_c) is given by Eq. (19).

$$T_c(s) = \frac{V_c}{V_c^*} = \frac{a_{2c}s^2 + a_{1c}s + a_{0c}}{C_f s^3 + a_{2c}s^2 + \left(C_f\omega_0^2 + a_{1c}\right)s + a_{0c}} \tag{19}$$

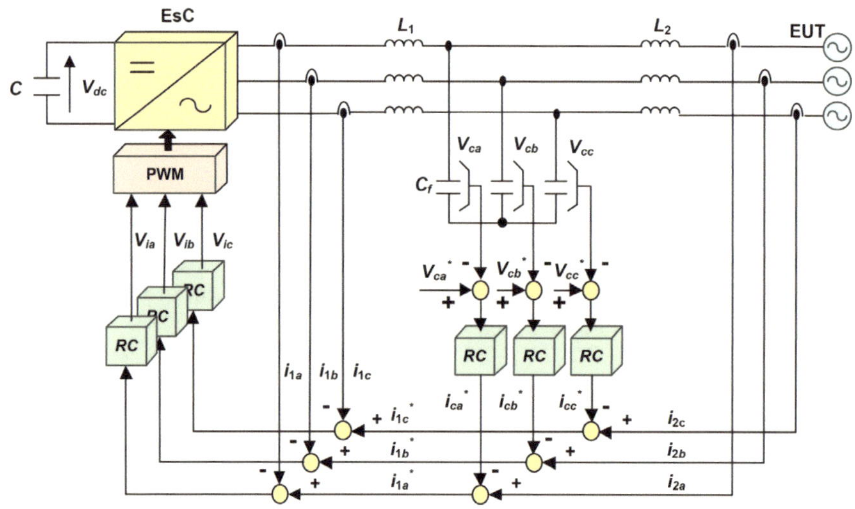

Figure 11.
Block diagram of the EsC control.

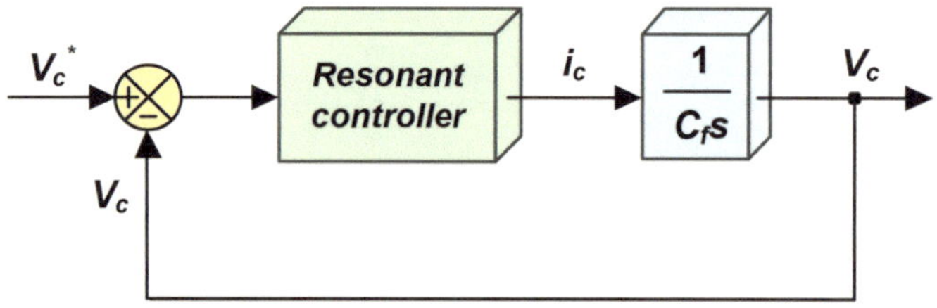

Figure 12.
Voltage regulation loop simplified block diagram.

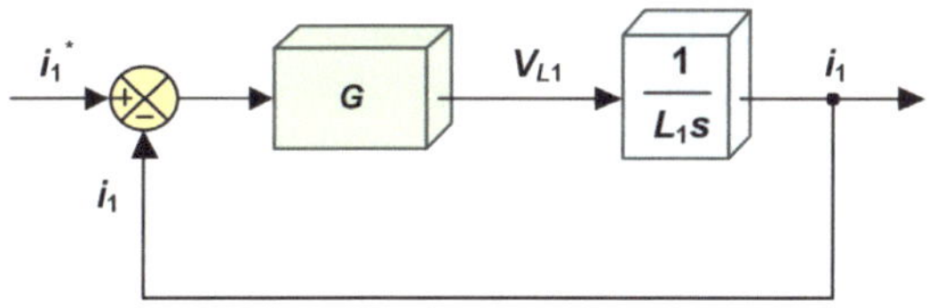

Figure 13.
Current regulation loop simplified block diagram.

The method chosen for the computation of the resonant controller parameters is based on the generalized stability criterion [22]. In this case, the n order polynomial is expressed as in Eq. (20).

$$P_{GSC}(s) = \lambda(s+r)\prod_{i=1}^{n}[(s+r+j\omega_i)(s+r-j\omega_i)]$$
$$\{\lambda, r, \omega_i \in \Re;\ i, n \in N\} \tag{20}$$

On the other hand, based on Eq. (19), the system characteristic polynomial P_c is given by Eq. (21).

$$P_c(s) = C_f s^3 + a_{2c}s^2 + \left(C_f\omega_0^2 + a_{1c}\right)s + a_{0c} \tag{21}$$

The identification of P_c and second order generalized stability criterion polynomial allows the deduction of the resonant controller parameters as shown in Eq. (22).

$$\begin{cases} a_{2c} = 3r_c\lambda_c \\ a_{1c} = \lambda_c\left(3r_c^2 + \omega_i^2\right) - C_f\omega_0^2 \\ a_{0c} = \lambda_c\left(r_c^3 + r_c\omega_i^2\right) \\ Avec\ \ \lambda_c = C_f \end{cases} \tag{22}$$

4.2.2 Tuning of the gain of the current i_1

The simplified internal current regulation loop block diagram is given by **Figure 13**.

Hence, the transfer function of the closed-loop system $T_{i1}(s)$ is given by Eq. (23).

$$T_{i1}(s) = \frac{i_1(s)}{i_1^*(s)} = \frac{1}{\frac{L}{G}s+1} = \frac{1}{1+\tau_c s} \quad \text{where } \tau_c = \frac{L_1}{G} \tag{23}$$

G is chosen so that the real part of the inverse of the closed-loop time constant $(1/\tau_c)$ is greater than the stability margin chosen for the synthesis of the voltage external loop in order to ensure that the internal loop is faster than the external one.

5. Line impedance emulation algorithms

In this section, two methods of the line impedance emulator algorithm synthesis are presented: the trigonometric functions-based algorithm and the voltage drop-based algorithm.

5.1 Trigonometric functions-based algorithm

The impedance emulation conception is based on the phasor diagram depicted on **Figure 14** According to this Figure, the apparent power S is expressed as in Eq. (24).

$$S = V_g I^* = V_g \left(\frac{V_g - V_c}{Z}\right)^* = \frac{V_g{}^2}{Z} e^{j\theta} - \frac{V_g V_c}{Z} e^{j(\theta+\delta)} \tag{24}$$

According to **Figure 14**, the reactive power Q and active power P are given by Eqs. (25) and (26), respectively. These equations allow the deduction of tanδ and the voltage magnitude V_{out} given, respectively, by Eqs. (27) and (28). On the other hand, the Q and P can be also written as a function of $\alpha\beta$ output current and voltage components as shown in Eqs. (29) and (30), respectively.

$$Q = \frac{V_g}{R^2 + X^2}\left[-RV_c \sin\delta + X\left(V_g - V_c \cos\delta\right)\right] \tag{25}$$

$$P = \frac{V_g}{R^2 + X^2}\left[R\left(V_g - V_c \cos\delta\right) + XV_c \sin\delta\right] \tag{26}$$

$$tan\delta = \frac{PX - QR}{V_g^2 - (PX + QR)} \tag{27}$$

$$V_c = \frac{PX - QR}{V_g \sin\delta} \tag{28}$$

$$Q = \frac{3}{2}\left(V_{c\beta} i_{2\alpha} + V_{c\alpha} i_{2\beta}\right) \tag{29}$$

$$P = \frac{3}{2}\left(V_{c\alpha} i_{2\alpha} + V_{c\beta} i_{2\beta}\right) \tag{30}$$

Figure 15 shows the trigonometric-based line impedance emulation algorithm. The first step consists in measuring the grid voltage $V_{g(a,b,c)}$ and computing its RMS value. From the obtained value, we compute the phase shifting δ relatively to the grid voltage. After that, the emulated impedance is computed based on the previous equations.

5.2 Voltage drop-based algorithm

This algorithm is based on a voltage drop V_v that matches with the emulated line impedance Z as shown in **Figure 16** This voltage is a function of programmed inductance and resistance variations as presented in Eq. (31). The voltage drop-based line impedance emulator algorithm is presented in **Figure 17**.

$$V_v = Zi_2 = (R + jX)i_2$$

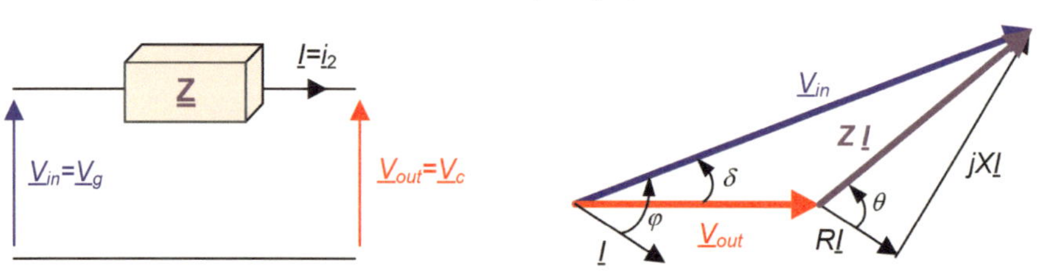

Figure 14.
Line impedance and phasor diagram.

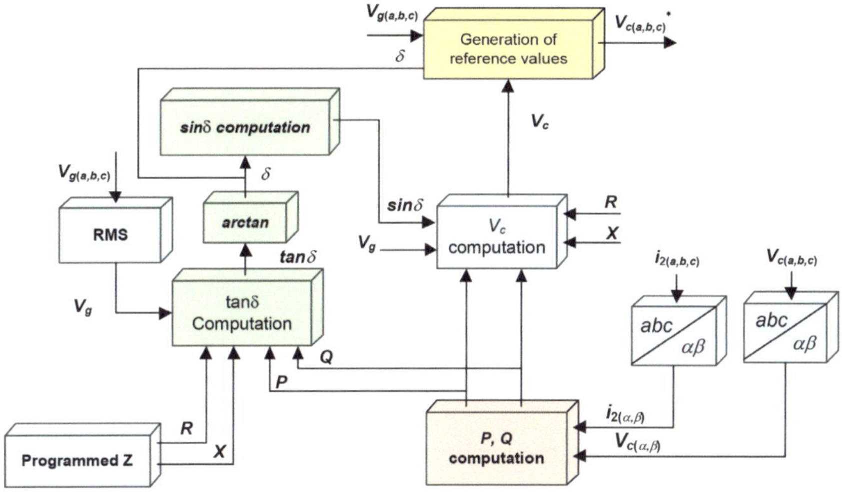

Figure 15.
Line impedance emulator algorithm-based trigonometric functions.

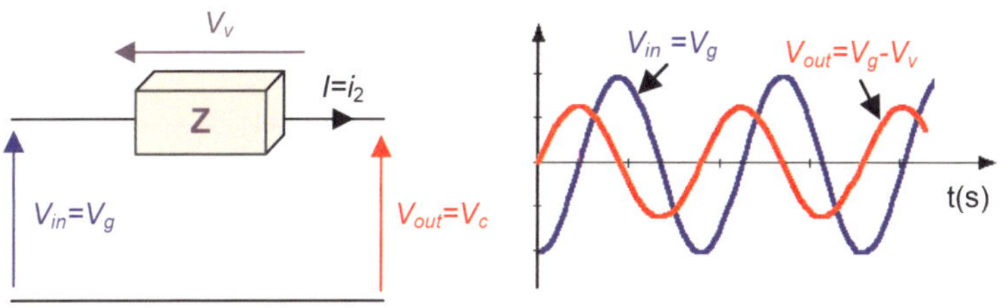

Figure 16.
Voltage drop line impedance emulator principle.

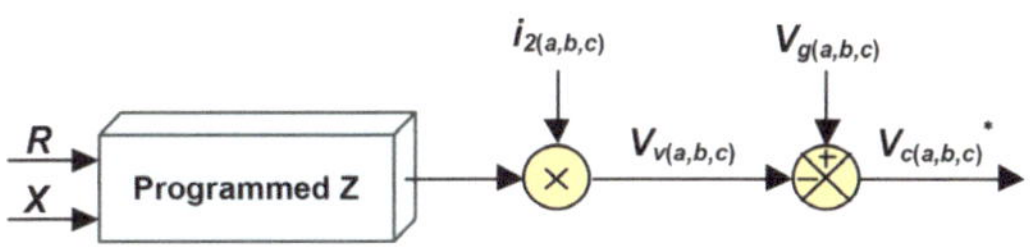

Figure 17.
Reference voltage according to fixed line impedance.

6. Simulation and discussion

Simulation tests were performed under PSIM software. The proposed control was applied to a 20kVA line impedance emulator. **Table 1** gives the line impedance emulator parameters. In **Figure 18** is presented the V_{dc} response to a step reference of 100 V. Based on this result, the steady state error of the V_{dc} voltage becomes null in the steady state, which prove that this voltage is well regulated. **Figure 19** shows that the voltage $V_{c(abc)}$ is well regulated in both transient and steady state operation even reference magnitude change at 0.9 s. To show the voltage drop-based line impedance emulation algorithm performances, a control scenario is presented in **Figure 20**. This scenario consists in imposing in the interval [0, 1 s] equivalent real impedance in series with L_2 and in the interval [1 s, 1.5 s] the line impedance emulator is activated. **Figure 21** shows results for a line impedance Z characterized by $X = 1.5\ \Omega$ and $R = 1\ \Omega$ in case of real and emulated impedance. As shown in this figure, the same current value is generated for real and programmed line impedances.

Description		Symbol	Value	Unit
Nominal voltage line-line		V_g	400	V
GsC nominal power		$SGsC_{nom}$	20	kVA
EsC nominal power		$SEsC_{nom}$	20	kVA
LCL filter	Converter side inductor	L_1	2	mH
	EUT side filter inductor	L_2	2	mH
	Capacitor	C_f	30	μF
Switching frequency		f_s	10	kHz

Table 1.
Line impedance emulator parameters.

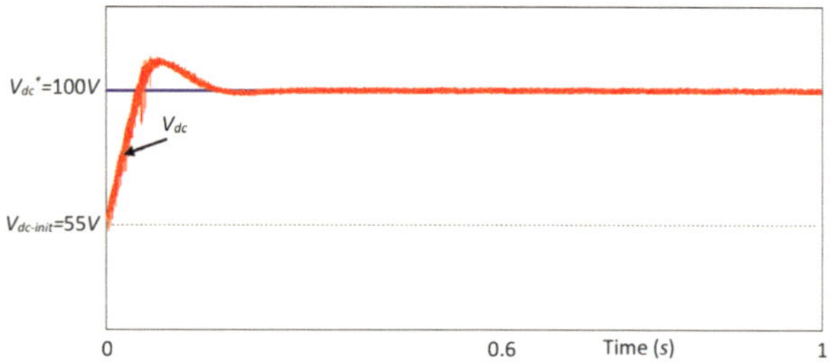

Figure 18.
V_{dc} response to a step reference of 100 V.

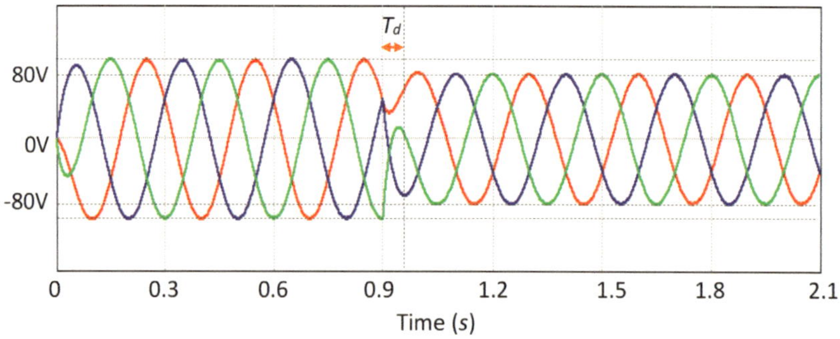

Figure 19.
Line impedance emulator output in case of voltage reference magnitude change.

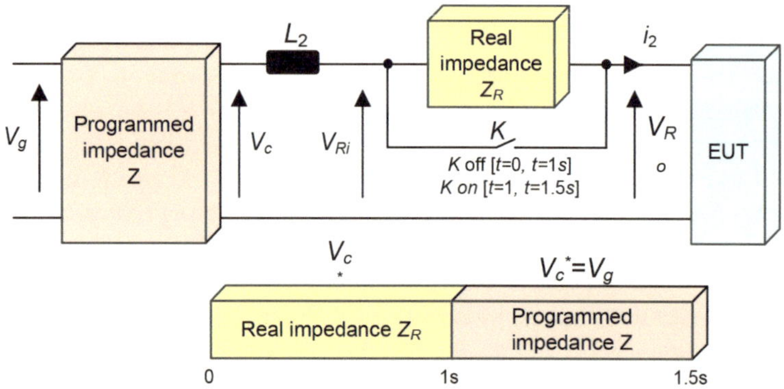

Figure 20.
Simulation control scenario.

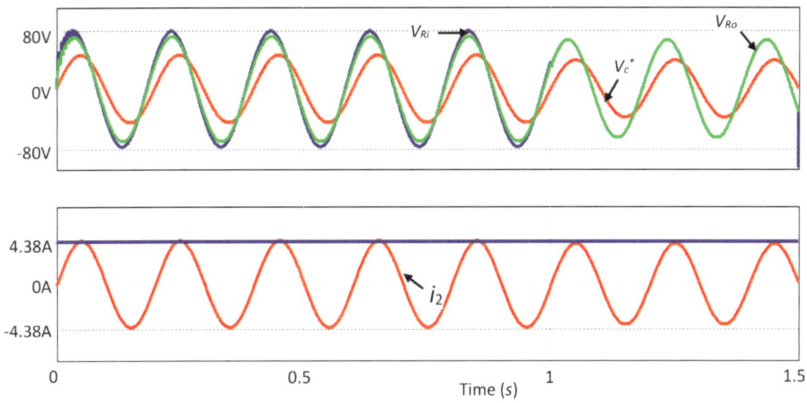

Figure 21.
System output for real and programmed impedance for X = 1.5 Ω and R = 1 Ω.

7. Experimental validation

Figures 22 and **23** show the experimental prototype and the test bench for the line impedance emulator. It includes (1) an auto transformer used in order to vary the voltage peak magnitude; (2) an L filter (composed of three inductors (20 mH/20A)

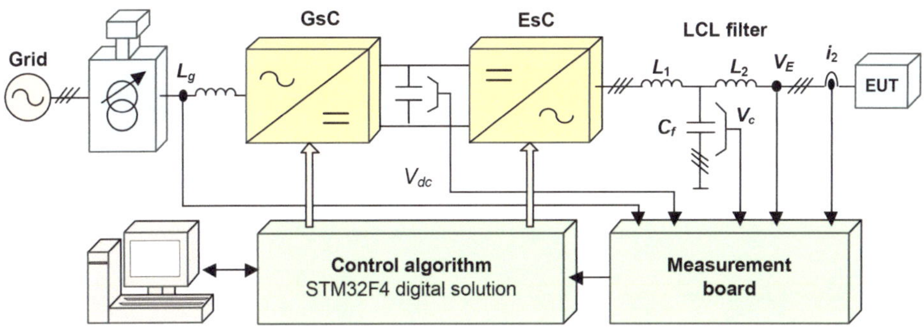

Figure 22.
Experimental prototype.

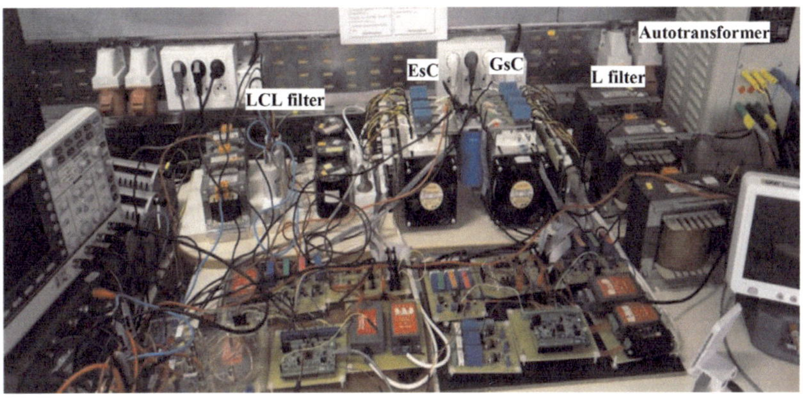

Figure 23.
Experimental test bench.

with 0.3 Ω internal resistors; (3) a 20 kVA AC/DC converter (GsC); (4) a dc-link capacitor (1100 μF/800 V); (5) a 20 kVA DC/AC converter (EsC); (6) an LCL filter (composed of three inductors (2 mH/10 A) with 0.1 Ω internal resistors, three capacitors (4 μF/400 V) and three inductors (2 mH/10A) with 0.1 Ω internal resistors); (7) a measurement board (LEM LA55 and LEM LV25 for currents and voltage measuring, respectively); and (8) the STM32F4-Discovery digital solution. It is worth noting here that two STM32F4-Discovery cards were used in the experimental test bench; the first one is dedicated to the GsC control and the second one is dedicated to the EsC control.

For both GsC and EsC controls, the switching frequency was fixed equal to 10 kHz. For experimental tests, the switching frequency is equal to 10 kHz, the voltage at the DC bus V_{dc} is initially charged at 55 V. **Figure 24** presents the voltage at the DC bus V_{dc} response. As shown is this figure, V_{dc} is well controlled during steady state operation. **Figure 25** presents the response of the line impedance emulator output for a reference change from 20 to 10 V. This test shows that the EsC control ensures an acceptable dynamic response and it is well controlled at steady state. **Figure 26** presents the line impedance emulator input and the output that matches with various values of line impedance.

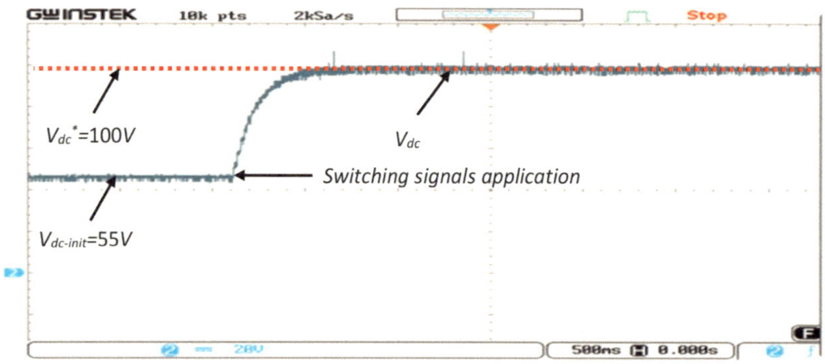

Figure 24.
DC bus measured voltage and reference values.

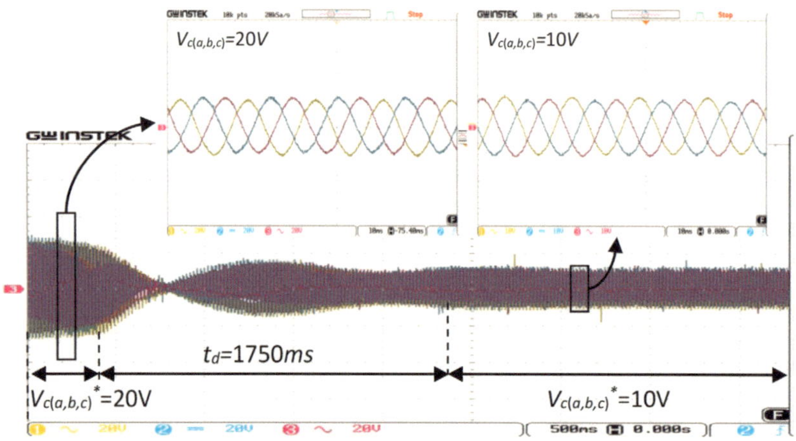

Figure 25.
Emulator output voltage $V_{c(abc)}$ for voltage reference change from 20 to 10 V.

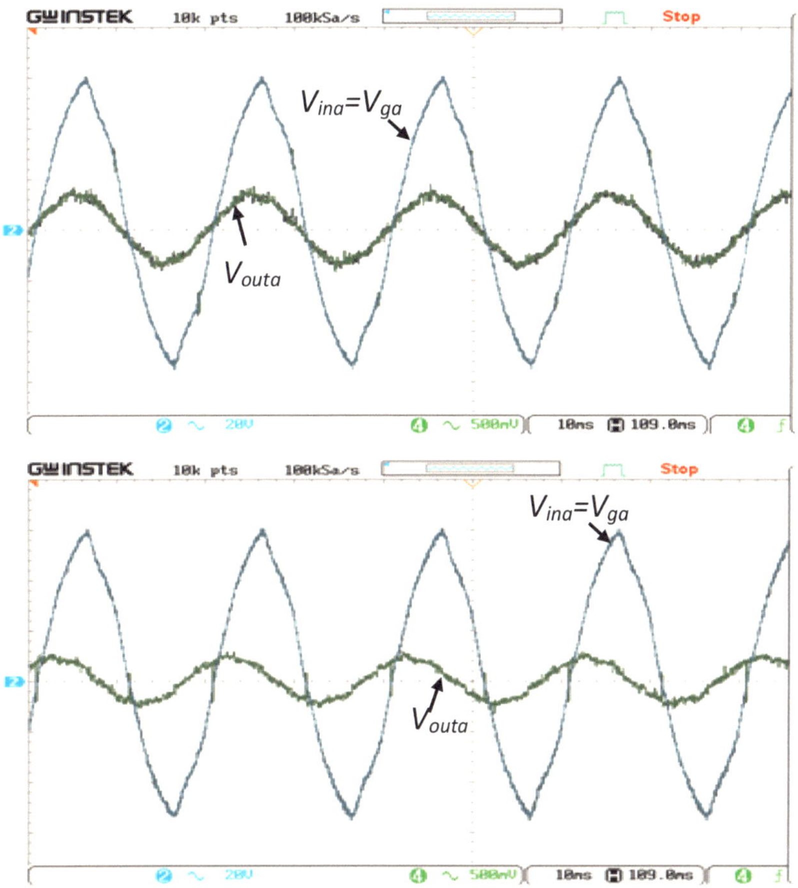

Figure 26.
Line impedance emulator input V_{ina} and output V_{outa} for different values of R and L.

8. Conclusion

In this chapter, line impedance emulator was studied. This equipment is used in small scale laboratories studying distributed energy generation. It ensures power tests with variable line impedance. Presented line impedance emulator is based on two power converters connected via a dc-link capacitor. Theoretical study is detailed and validated by simulation and experimental tests. The proposed study describes in detail the control design of each power converter. In addition, two variants of line impedance emulator algorithms were synthesized. To prove the efficiency of the presented study, a test with a real impedance and an emulated one was performed and obtained results show the similarity of system responses with both equipment.

Acknowledgements

This work was supported by the Tunisian Ministry of High Education and Research under Grant LSE-ENIT-LR 11ES15.

Author details

Marwa Ben Saïd-Romdhane[1*], Sondes Skander-Mustapha[1,2] and Ilhem Slama-Belkhodja[1]

1 Université de Tunis El Manar, Ecole Nationale d'Ingénieurs de Tunis, Laboratoire des Systèmes Electriques, LR11ES15, Tunis, Tunisie

2 Université de Carthage, Ecole Nationale d'Architecture et d'Urbanisme, Sidi Bou Said, Tunisie

*Address all correspondence to: marwa.bensaidromdhane@enit.utm.tn

IntechOpen

© 2020 The Author(s). Licensee IntechOpen. This chapter is distributed under the terms of the Creative Commons Attribution License (http://creativecommons.org/licenses/by/3.0), which permits unrestricted use, distribution, and reproduction in any medium, provided the original work is properly cited.

References

[1] Ahuja H, Kumar P. A novel approach for coordinated operation of variable speed wind energy conversion in smart grid applications. Computers & Electrical Engineering. 2019;**77**:72-87. DOI: 10.1016/j.compeleceng.2019.05.004

[2] Bollen MHJ. Power quality concerns in implementing smart distribution-grid applications. IEEE Transactions on Smart Grid. 2017;**8**:391-399. DOI: 10.1109/TSG.2016.2596788

[3] Titmus P, Strickland D, Cross A. Low cost laboratory micro-grid hardware and control for electrical power systems teaching. In: 2017 19th European Conference on Power Electronics and Applications, EPE 2017 ECCE Europe, 2017–January, 1–10. 2017. DOI: 10.23919/EPE17ECCEEurope.2017.8098945

[4] Twaha S, Ramli MAM. A review of optimization approaches for hybrid distributed energy generation systems: Off-grid and grid-connected systems. Sustainable Cities and Society. 2018. DOI: 10.1016/j.scs.2018.05.027

[5] Skander-Mustapha S, Jebali-Ben Ghorbal M, Said-Romdhane MB, Miladi M, Slama-Belkhodja I. Grid emulator for small scale distributed energy generation laboratory. Sustainable Cities and Society. 2018;**43**: 325-338. DOI: 10.1016/j.scs.2018.09.007

[6] Skander-Mustapha S, Ghorbal MJ, Miladi M, Slama-Belkhodja I. Load analysis effect on grid fault emulator. International Renewable Energy Congress. 2018:1-6. DOI: 10.1109/IREC.2018.8362551

[7] Said-Romdhane MB, Skander-Mustapha S, Slama-Belkhodja I. Enhanced real time impedance emulation for microgrid equipments testing and applications. International Renewable Energy Congress. 2019:1-6. DOI: 10.1109/IREC.2019.8754577

[8] Espinoza JL, Gonzalez LG, Sempertegui R. Micro grid laboratory as a tool for research on non-conventional energy sources in Ecuador. In: 2017 IEEE International Autumn Meeting on Power, Electronics and Computing, ROPEC 2017, 2018–January (ROPEC). 2018. pp. 1-7. DOI: 10.1109/ROPEC.2017.8261615

[9] Patrascu C, Muntean N, Cornea O, Hedes A. Microgrid laboratory for educational and research purposes. In: IEEE 16th International Conference on Environment and Electrical Engineering (EEEIC). 2016

[10] Requirements for the Connection of Micro-Generators in Parallel with Public Low Voltage Distribution Networks. European Standard EN 50438; 2007

[11] Swiss Grid. Transmission Code 2013. [Online]. 2013. Available at: https://www.strom.ch/fileadmin/user_upload/Dokumente_Bilder_neu/010_Downloads/Branchenempfehlung/Document_de_la_branche_TC_2013.pdf

[12] Nordel. Nordic Grid Code 2007 edition. 2007 [online]. Available at: https://www.entsoe.eu/fileadmin/user-upload/-library/publications/nordic/planning/070115-entsoe-nordic-NordicGridCode.pdf

[13] E. On Netz, Grid Code: High and Extra High Voltage. 2006 [Online]. Available at: http://www.eonnetz.com/pages/ehn-de/Veroeffentlichungen/Netzanschluss/Netzanschlussregeln/ENENARHS2006eng.pdf

[14] Timbus AV, Rodriguez P, Teodorescu R, Ciobotaru M. Line impedance estimation using active and reactive power variations. 2007 IEEE Power Electronics Specialists

Conference, Orlando, FL; 2007. pp. 1273-1279. DOI: 10.1109/PESC.2007.4342176

[15] Tuladhar A, Hua J, Unger T, Mauch K. Control of parallel inverters in distributed AC power systems with consideration of line impedance effect. IEEE Transactions on Industry Applications. Jan.-Feb. 2000;**36**(1): 131-138. DOI: 10.1109/28.821807

[16] Ye P, He J, Wang G, Li S, Sun F, Han Y, et al. An improved droop control strategy for parallel inverters in microgrid. In: IEEE Conference on Energy Internet and Energy System Integration (EI2). 2017. DOI: 10.1109/EI2.2017.8245514

[17] Raj D, Chethan, Gaonkar DN, Guerrero JM. Improved P-f/Q-V and P-V/Q-f droop controllers for parallel distributed generation inverters in AC microgrid. Sustainable Cities and Society. 2018. DOI: 10.1016/j.scs.2018.04.026

[18] He J, Du L, Liang B, Li Y, Wang C. A coupled virtual impedance for parallel AC/DC converter based power electronics system. IEEE Transactions on Smart Grid. May 2019;**10**(3): 3387-3400. DOI: 10.1109/TSG.2018.2825383

[19] Phan VT, Lee HH. Control strategy for harmonic elimination in stand-alone DFIG applications with nonlinear loads. IEEE Transactions on Power Electronics. 2011;**26**(9):2662-2675

[20] Bacca S, Munteanu I, Bratcu AI. Power Electronic Converters Modeling and Control. London, England: Springer-Verlag; 2014:257-261

[21] Hautier JP, Guillaud X, Vandecasteele F, Wulveryck M. Contrôle de grandeurs alternatives par correcteur résonant. Revue Internationale de Génie Électrique. 1999;**2**:163-183